CLEP* CHEMISTRY

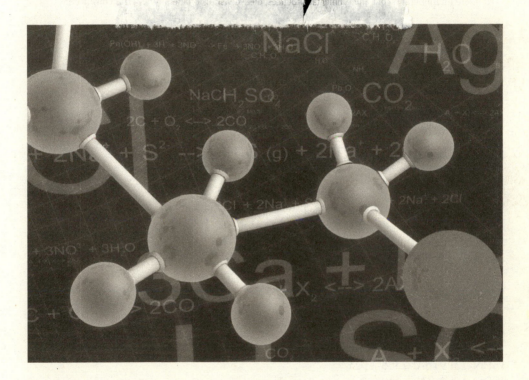

Kevin Reel

Head of School
The Colorado Springs School
Colorado Springs, Colorado

D0073906

Research & Education Association
Visit our website at: www.rea.com

Planet Friendly Publishing
✔ Made in the United States
✔ Printed on Recycled Paper
GREEN EDITION®   Text: 10%   Cover: 10%
Learn more: www.greenedition.org

At REA we're committed to producing books in an Earth-friendly manner and to helping our customers make greener choices.

Manufacturing books in the United States ensures compliance with strict environmental laws and eliminates the need for international freight shipping, a major contributor to global air pollution.

And printing on recycled paper helps minimize our consumption of trees, water and fossil fuels. This book was printed on paper made with **10% post-consumer waste**. According to the Environmental Paper Network's Paper Calculator, by using this innovative paper instead of conventional papers, we achieved the following environmental benefits:

Trees Saved: 3 • Air Emissions Eliminated: 563 pounds
Water Saved: 571 gallons • Solid Waste Eliminated: 166 pounds

Courier Corporation, the manufacturer of this book, owns the Green Edition Trademark. For more information on our environmental practices, please visit us online at **www.rea.com/green**

Research & Education Association
61 Ethel Road West
Piscataway, New Jersey 08854
E-mail: info@rea.com

CLEP Chemistry with Online Practice Exams

Published 2014
Copyright © 2013 by Research & Education Association, Inc. Prior edition copyright © 2006 by Research & Education Association, Inc. All rights reserved. No part of this book may be reproduced in any form without permission of the publisher.

Printed in the United States of America

Library of Congress Control Number: 2012947519

ISBN-13: 978-0-7386-1103-7
ISBN-10: 0-7386-1103-4

REA® is a registered trademark of Research & Education Association, Inc.

CONTENTS

CHAPTER 4

CHAPTER 5

CHAPTER 6

CHAPTER 7

CHAPTER 8

ABOUT RESEARCH & EDUCATION ASSOCIATION

Founded in 1959, Research & Education Association (REA) is dedicated to publishing the finest and most effective educational materials—including study guides and test preps—for students in middle school, high school, college, graduate school, and beyond.

Today, REA's wide-ranging catalog is a leading resource for teachers, students, and professionals. Visit *www.rea.com* to see a complete listing of all our titles.

ACKNOWLEDGMENTS

We would like to thank Pam Weston, Publisher, for setting the quality standards for production integrity and managing the publication to completion; John Paul Cording, Vice President, Technology, for coordinating the design and development of the online REA Study Center; Larry B. Kling, Vice President, Editorial, for his supervision of revisions and overall direction; Diane Goldschmidt and Michael Reynolds, Managing Editors, for coordinating development of this edition; Patrica Van Arnum for technically reviewing the manuscript; Transcend Creative Services for typesetting this edition; and Weymouth Design and Christine Saul for designing our cover.

CHAPTER 1

Passing the CLEP Chemistry Exam

PASSING THE CLEP CHEMISTRY EXAM

Congratulations! You're joining the millions of people who have discovered the value and educational advantage offered by the College Board's College-Level Examination Program, or CLEP. This test prep covers everything you need to know about the CLEP Chemistry exam, and will help you earn the college credit you deserve while reducing your tuition costs.

GETTING STARTED

There are many different ways to prepare for a CLEP exam. What's best for you depends on how much time you have to study and how comfortable you are with the subject matter. To score your highest, you need a system that can be customized to fit you: your schedule, your learning style, and your current level of knowledge.

This book, and the online tools that come with it, allow you to create a personalized study plan through three simple steps: assessment of your knowledge, targeted review of exam content, and reinforcement in the areas where you need the most help.

Let's get started and see how this system works.

Test Yourself and Get Feedback	Assess your strengths and weaknesses. The score report from your online diagnostic exam gives you a fast way to pinpoint what you already know and where you need to spend more time studying.
Review with the Book	Armed with your diagnostic score report, review the parts of the book where you're weak and study the answer explanations for the test questions you answered incorrectly.
Ensure You're Ready for Test Day	After you've finished reviewing with the book, take our full-length practice tests. Review your score reports and re-study any topics you missed. We give you two full-length practice tests to ensure you're confident and ready for test day.

THE REA STUDY CENTER

The best way to personalize your study plan and focus on your weaknesses is to get feedback on what you know and what you don't know. At the online REA Study Center, you can access two types of assessment: a diagnostic exam and full-length practice exams. Each of these tools provides true-to-format questions and delivers a detailed score report that follows the topics set by the College Board.

Diagnostic Exam

Before you begin your review with the book, take the online diagnostic exam. Use your score report to help evaluate your overall understanding of the subject, so you can focus your study on the topics where you need the most review.

Full-Length Practice Exams

These practice tests give you the most complete picture of your strengths and weaknesses. After you've finished reviewing with the book, test what you've learned by taking the first of the two online practice exams. Review your score report, then go back and study any topics you missed. Take the second practice test to ensure you have mastered the material and are ready for test day.

If you're studying and don't have Internet access, you can take the printed tests in the book. These are the same practice tests offered at the REA Study Center, but without the added benefits of timed testing conditions and diagnostic score reports. Because the actual exam is computer-based, we recommend you take at least one practice test online to simulate test-day conditions.

AN OVERVIEW OF THE EXAM

The CLEP Chemistry exam consists of 75 multiple-choice questions, each with five possible answer choices, to be answered in 90 minutes.

The exam covers the material one would find in a one-year college-level general chemistry course. The exam emphasizes the following topics: structure and states of matter, reaction types, equations and stoichiometry, equilibrium, kinetics, thermodynamics, and descriptive and experimental chemistry.

The approximate breakdown of topics is as follows:

20%	Structure of Matter
19%	States of Matter
12%	Reaction Types
10%	Equations and Stoichiometry
7%	Equilibrium
4%	Kinetics
5%	Thermodynamics
14%	Descriptive Chemistry
9%	Experimental Chemistry

ALL ABOUT THE CLEP PROGRAM

What is the CLEP?

CLEP is the most widely accepted credit-by-examination program in North America. CLEP exams are available in 33 subjects and test the material commonly required in an introductory-level college course. Examinees can earn from three to twelve credits at more than 2,900 colleges and universities in the U.S. and Canada. For a complete list of the CLEP subject examinations offered, visit the College Board website: *www.collegeboard.org/clep*.

Who takes CLEP exams?

CLEP exams are typically taken by people who have acquired knowledge outside the classroom and who wish to bypass certain college courses and earn college credit. The CLEP program is designed to reward examinees for learning—no matter where or how that knowledge was acquired.

Although most CLEP examinees are adults returning to college, many graduating high school seniors, enrolled college students, military personnel, veterans, and international students take CLEP exams to earn college credit or to demonstrate their ability to perform at the college level. There are no prerequisites, such as age or educational status, for taking CLEP examinations. However, because policies on granting credits vary among colleges, you should contact the particular institution from which you wish to receive CLEP credit.

Who administers the exam?

CLEP exams are developed by the College Board, administered by Educational Testing Service (ETS), and involve the assistance of educators from throughout the United States. The test development process is designed and implemented to ensure that the content and difficulty level of the test are appropriate.

When and where is the exam given?

CLEP exams are administered year-round at more than 1,200 test centers in the United States and can be arranged for candidates abroad on request. To find the test center nearest you and to register for the exam, contact the CLEP Program:

CLEP Services
P.O. Box 6600
Princeton, NJ 08541-6600
Phone: (800) 257-9558 (8 A.M. to 6 P.M. ET)
Fax: (609) 771-7088
Website: *www.collegeboard.org/clep*

CLEP MIGRATES TO INTERNET-BASED TESTING

To improve the testing experience for institutions and test-takers, the CLEP program is transitioning its 33 exams from the current eCBT platform to the more sophisticated Internet-based testing (iBT) platform. Test-takers will be able to register for exams and manage their personal account information through the My Account feature on the CLEP website. This new feature will simplify the registration activities that need to be performed by test center administrators (TCAs). iBT automatically downloads all pertinent information about the test session from My Account. The check-in process will be more streamlined than in eCBT. There is no longer a need for TCAs to collect profile or payment information as is necessary in eCBT.

Key Benefits of Using iBT

Here are some of the many benefits you will enjoy using the new iBT platform:

- Sophisticated testing platform with contemporary design
- Stable performance with reduced software challenges
- Simplified TCA check-in process
- Reduced time for administrative activities on test day
- Test platform integrated with the CLEP website
- Automated software distribution to replace shipment of CDs

For detailed information on the process of transitioning from eCBT to iBT, download the brochure **Announcing CLEP's Migration to iBT**. Also, please refer to the FAQs page and download the **Test Delivery Requirements**.

OPTIONS FOR MILITARY PERSONNEL AND VETERANS

CLEP exams are available free of charge to eligible military personnel and eligible civilian employees. All the CLEP exams are available at test centers on college campuses and military bases. Contact your Educational Services Officer or Navy College Education Specialist for more information. Visit the DANTES or College Board websites for details about CLEP opportunities for military personnel.

Eligible U.S. veterans can claim reimbursement for CLEP exams and administration fees pursuant to provisions of the Veterans Benefits Improvement Act of 2004. For details on eligibility and submitting a claim for reimbursement, visit the U.S. Department of Veterans Affairs website at *www.gibill.va.gov.*

CLEP can be used in conjunction with the Post-9/11 GI Bill, which applies to veterans returning from the Iraq and Afghanistan theaters of operation. Because the GI Bill provides tuition for up to 36 months, earning college credits with CLEP exams expedites academic progress and degree completion within the funded timeframe.

SSD ACCOMMODATIONS FOR CANDIDATES WITH DISABILITIES

Many test candidates qualify for extra time to take the CLEP exams, but you must make these arrangements in advance. For information, contact:

College Board Services for Students with Disabilities
P.O. Box 6226
Princeton, NJ 08541-6226
Phone: (609) 771-7137 (Monday through Friday, 8 A.M. to 6 P.M. ET)
TTY: (609) 882-4118
Fax: (609) 771-7944
E-mail: ssd@info.collegeboard.org

6-WEEK STUDY PLAN

Although our study plan is designed to be used in the six weeks before your exam, it can be condensed to three weeks by combining each two-week period into one.

Be sure to set aside enough time—at least two hours each day—to study. The more time you spend studying, the more prepared and relaxed you will feel on the day of the exam.

Week	Activity
1	Take the Diagnostic Exam at the online REA Study Center. Your score report will identify topics where you need the most review.
2–4	Study the review, focusing on the topics you missed (or were unsure of) on the Diagnostic Exam.
5	Take Practice Test 1 at the REA Study Center. Review your score report and re-study any topics you missed.
6	Take Practice Test 2 at the REA Study Center to see how much your score has improved. If you still got a few questions wrong, go back to the review and study the topics you missed.

TEST-TAKING TIPS

Know the format of the test. CLEP computer-based tests are fixed-length tests. This makes them similar to the paper-and-pencil type of exam because you have the flexibility to go back and review your work in each section.

Learn the test structure, the time allotted for each section of the test, and the directions for each section. By learning this, you will know what is expected of you on test day, and you'll relieve your test anxiety.

Read all the questions—completely. Make sure you understand each question before looking for the right answer. Reread the question if it doesn't make sense.

Read all of the answers to a question. Just because you think you found the correct response right away, do not assume that it's the best answer. The last answer choice might be the correct answer.

Work quickly and steadily. You will have 90 minutes to answer 75 questions, so work quickly and steadily. Taking the timed practice tests online will help you learn how to budget your time.

Use the process of elimination. Stumped by a question? Don't make a random guess. Eliminate as many of the answer choices as possible. By eliminating just two answer choices, you give yourself a better chance of getting the item correct, since there will only be three choices left from which to make your guess. Remember, your score is based only on the number of questions you answer correctly.

Don't waste time! Don't spend too much time on any one question. Remember, your time is limited and pacing yourself is very important. Work on the easier questions first. Skip the difficult questions and go back to them if you have the time.

Look for clues to answers in other questions. If you skip a question you don't know the answer to, you might find a clue to the answer elsewhere on the test.

Acquaint yourself with the computer screen. Familiarize yourself with the CLEP computer screen beforehand by logging on to the College Board website. Waiting until test day to see what it looks like in the pretest tutorial risks injecting needless anxiety into your testing experience. Also, familiarizing yourself with the directions and format of the exam will save you valuable time on the day of the actual test.

Be sure that your answer registers before you go to the next item. Look at the screen to see that your mouse-click causes the pointer to darken the proper oval. If your answer doesn't register, you won't get credit for that question.

THE DAY OF THE EXAM

On test day, you should wake up early (after a good night's rest, of course) and have breakfast. Dress comfortably, so you are not distracted by being too hot or too cold while taking the test. (Note that "hoodies" are not allowed.) Arrive at the test center early. This will allow you to collect your thoughts and relax before the test, and it will also spare you the anxiety that comes with being late. As an added incentive, keep in mind that no one will be allowed into the test session after the test has begun.

Before you leave for the test center, make sure you have your admission form and another form of identification, which must contain a recent photograph, your name, and signature (i.e., driver's license, student identification card, or current alien registration card). You will not be admitted to the test center if you do not have proper identification.

You may wear a watch to the test center. However, you may not wear one that makes noise, because it may disturb the other test-takers. No cell phones, dictionaries, textbooks, notebooks, briefcases, or packages will be permitted, and drinking, smoking, and eating are prohibited.

Good luck on the CLEP Chemistry exam!

CHAPTER 2

The Structure of Matter

THE STRUCTURE OF MATTER

ATOMIC THEORY AND STRUCTURE

EVIDENCE FOR ATOMIC THEORY

Dalton

- John Dalton performed chemical reactions and carefully measured the masses of reactants and products.
- Dalton proposed that all matter is composed of subunits call **atoms**. Atoms had different identities, called **elements**. Elements combined together in definite ratios to form **compounds**.
- Atoms are never created or destroyed during chemical reactions.

Thompson

- J. J. Thompson observed the deflection of particles in a cathode ray tube.
- He concluded that atoms are composed of positive and negative charges.
- He called negative charges **electrons**, and he suggested that the positive charges were distributed in islands throughout the atom, like raisins in raisin bread. Some people at the time called this the "plum pudding" model of the atom.

Millikan

- Robert Millikan used oil drops falling in an electric field of known strength to calculate the charge-to-mass ratio of electrons and to surmise the charge contained by a single electron.

Rutherford

- Ernest Rutherford fired alpha particles (that he knew to be positively charged) through thin, gold foil.

- Rutherford measured the resulting scatter patterns of the alpha particles after they hit the foil. He found that most of the alpha particles moved right through the foil, or were deflected slightly. However, some alpha particles were deflected at large angles, as though they had collided with a heavier object and bounced back.

- He concluded that the positive charge and the mass of the atom are concentrated at the center of the atom, and that the rest of the atom is mostly empty space. This directly countered Thompson's "plum pudding" model.

Planck

- Max Planck determined that electromagnetic energy is quantized, or composed of discrete bundles, expressed by the equation below.

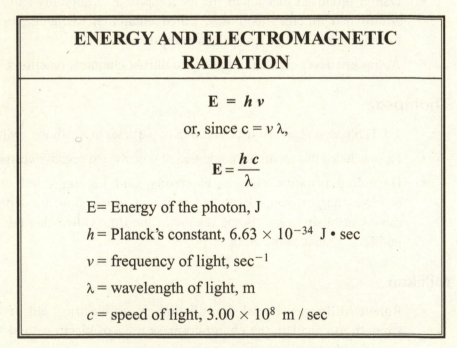

ENERGY AND ELECTROMAGNETIC RADIATION

$$E = h\,v$$

or, since $c = v\,\lambda$,

$$E = \frac{h\,c}{\lambda}$$

E = Energy of the photon, J

h = Planck's constant, 6.63×10^{-34} J • sec

v = frequency of light, sec^{-1}

λ = wavelength of light, m

c = speed of light, 3.00×10^8 m / sec

Bohr

- Niels Bohr applied the idea of quantized energy to show that electrons exist around the nucleus at a fixed radius.

- Electrons with higher energy exist farther from the nucleus.

- The Bohr model is accurate only for atoms and ions with one electron. It was clear that a more complex model was needed to explain atoms with multiple electrons.

- Electrons give off energy in the form of electromagnetic radiation when they move from a higher level, or an excited state, to a lower level. The energy represented by light, using Planck's equation, represents the difference between the two energy levels of the electron.

de Broglie

- Louis de Broglie identified the wave characteristics of matter by combining Einstein's relationship between mass and energy ($E = mc^2$) and the relationship between velocity and the wavelength of light ($E = hv$).

- This shows that all particles with momentum have a corresponding wave nature.

---

de BROGLIE WAVELENGTH OF PARTICLES

$$\lambda = \frac{h}{mv}$$

λ = wavelength associated with particle, m

m = mass of particle, kg

h = Planck's constant, 6.63×10^{-34} J • sec

v = velocity of particle, m/sec

mv = momentum of particle, kg m /sec

Heisenberg

- Werner Heisenberg, in the early 20th century, said that it is impossible to simultaneously know both the position and the momentum of an electron.

- For small particles, such as electrons, this uncertainty suggests that we need a wave model, rather than a Newtonian model, to understand their behavior.

Schrödinger

- Erwin Schrödinger, in the early 20th century, attributed a wave function to electrons.

- The wave function describes the probability of where an electron might exist. The regions of high probability are called **orbitals**, even though they are more like clouds than orbits.

- The orbital of each electron is described in Schrödinger's equation. These orbitals can be described as s, p, d, or f orbitals, as used in electron configurations described later in this chapter.

ATOMIC MASS, ATOMIC NUMBER, ISOTOPES

Atomic Mass

- The mass of an atom consists of the cumulative mass of all the particles in the atom, which includes protons, neutrons, and electrons.

- The mass of the electrons is insignificant relative to the mass of protons or neutrons. Therefore, the atomic mass is calculated by adding up the masses of the protons and neutrons.

- Example:

  A helium atom consists of two protons and two electrons. It would have an atomic mass of 4 amu. (atomic mass units, or the mass of one proton or neutron). It is depicted as helium-4.

Atomic Number

- The atomic number is composed of the total nuclear charge, or the number of protons in the nucleus of the atom.

- The atomic number is also the number of electrons surrounding the nucleus of a neutral atom.
- The atomic number is the smaller of the two numbers that exist for each element on the periodic table.
- Example:

  An atom of the element carbon-14 has an atomic number of 6. Therefore, when it is neutral, it has 6 protons in its nucleus, and 6 electrons around the nucleus.

Isotopes

- Atoms with the same number of protons but with different numbers of neutrons are **isotopes** of one another.
- Example:

  Carbon-12 and carbon-14 are the same element (carbon), which is defined as having 6 protons. The difference in the mass numbers indicates that carbon-12 has 6 neutrons, while carbon-14 has 8 neutrons.

Average Mass Number

- In nature, elements naturally occur as a combination of more than one isotope.
- The average mass number takes into account the relative frequencies of the different isotopes.
- The average mass number is also called the **atomic weight**.
- This number is also the **molar mass** of the element, or the mass in grams of one mole of atoms. 6.02×10^{23} atoms constitutes one mole of atoms.
- This is the larger number that exists on the periodic table of elements.
- The atomic weight is probably closest in value to the most commonly existing isotope.

  Example:

  Naturally occurring lead (Pb) exists as a combination of four isotopes, Pb-204, Pb-206, Pb-207, and Pb-208. Given the natural abundance of each isotope described in the following chart, calculate the average mass number of lead.

Pb-204	1.42%
Pb-206	24.10%
Pb-207	22.10%
Pb-208	52.40%

Solution:

Multiply the frequency, expressed as a fraction, by the mass of each isotope. Then add the contributions of each isotope together to get the total average mass.

$$(0.0142)(204) + (0.241)(206) + (0.221)(207) + (0.524)(208) = \underline{\mathbf{207.3}}$$

ELECTRONS

- Schrödinger's equation describes the location and shape taken by each **electron cloud** in an atom.

- Each electron can be described by a set of four quantum numbers. The **Pauli Exclusion Principle** states that no two electrons can occupy the same exact energy level or have the same set of four quantum numbers.

- There are three ways to model the location of electrons in atoms: quantum numbers, orbital notation, and electron configurations.

- The **quantum numbers** are like an address that describes the general distance from the nucleus, the type of orbital filled, the orientation of that orbital, and the "spin" direction on each electron in each orbital.

- **Orbital notation** identifies where each electron exists in each orbital. In this model, it is clear whether or not electrons have parallel spin.

- **Electron configurations** identify the number of electrons in each type of orbital at each energy level.

- According to the **Aufbau Principle**, electrons exist first at the lowest possible energy level, unless energy has put them into an excited state.

- **Hund's Rule** states that electrons enter orbitals of equal energy singly, with the same spin (*i.e.*, unpaired), before they become paired.

Quantum Numbers

Every electron in an atom can be uniquely described with a different combination of the four quantum numbers described below. Each combination of quantum numbers describes a unique level of energy contained by the electron. Each of the four numbers represents a different trait of the electron.

Principal Quantum Number: n

- The **principal quantum number** represents the shell an electron occupies.
- Shells are also called **energy levels**.
- Shells can have possible values of n = 1,2...7.
- This value indicates the approximate distance to the nucleus and relative energy. Therefore, electrons with higher values of n are farther from the nucleus and have higher energy.

Angular Momentum Quantum Number: l

- **Angular momentum quantum number** represents the subshell the electron occupies.
- This number describes the shape of an electron's orbital. Possible shapes include the following:

    When n = 1, l = 0 (meaning there is one possible type of orbital, s)

    When n = 2, l = 0 (s orbital) or l = 1 (p orbital)

    When n = 3, l = 0 (s orbital), l = 1 (p orbital), or l = 2 (d orbital)

    When n ≥ 4, l = 0 (s orbital), l = 1 (p orbital), l = 2 (d orbital), or l = 3 (f orbital)

Magnetic Quantum Number: m_l

- **Magnetic quantum number** represents the orbital position.
- Possible values range from –1...0...+1 for all possible values of n.

    When l = 0, m_l = 0 (there is one value, representing one possible s orbital)

    When l = 1, m_l = –1, 0, 1 (there are three possible p orbitals)

    When l = 2, m_l = –2, –1, 0, 1, 2 (there are five possible d orbitals)

    When l = 3, m_l = –3, –2, –1, 0, 1, 2, 3 (there are seven possible f orbitals)

- The orbital with the most negative number is filled first.

Magnetic Spin Quantum Number: m_s

- Each orbital can contain as many as two electrons, one with a positive spin (+1/2) and one with a negative spin (–1/2).
- The first ground state electron in each orbital takes the +1/2 value.

Atomic Orbitals and Electron Configurations

Table — Electron Arrangements

	Main Levels n = 1	n = 2				n = 3	Summary
	Sublevels s ($l = 0$)	s ($l = 0$)	p ($l = 1$)			s ($l = 0$)	
H	↑						$1s^1$
He	↑↓						$1s^2$
Li	↑↓	↑					$1s^22s^1$
Be	↑↓	↑↓					$1s^22s^2$
B	↑↓	↑↓	↑	○	○		$1s^22s^22p^1$
C	↑↓	↑↓	↑	↑	○		$1s^22s^22p^2$
N	↑↓	↑↓	↑	↑	↑		$1s^22s^22p^3$
O	↑↓	↑↓	↑↓	↑	↑		$1s^22s^22p^4$
F	↑↓	↑↓	↑↓	↑↓	↑		$1s^22s^22p^5$
Ne	↑↓	↑↓	↑↓	↑↓	↑↓		$1s^22s^22p^6$

Paramagnetism and Dimagnetism

- **Dimagnetic** elements have paired electrons in each orbital. To have this situation, all subshells are filled. These elements are not affected by magnetic fields.

- **Paramagnetic** elements have an unpaired electron in at least one orbital. The unpaired electron creates a magnetic field in the atom that responds to external magnetic fields.

- **Molecular orbitals** also create dimagnetism and paramagnetism.

PERIODIC TRENDS

Atomic Radii

- Moving from left to right across a period, the atomic radius decreases.
- Moving down a group from top to bottom, the atomic radius increases.
- Cations (positive ions) have smaller radii than their corresponding neutral atoms.
- Anions (negative ions) have larger radii than their corresponding neutral atoms.

Ionization Energy

- Moving from left to right across a period, ionization energy increases.
- Moving down a group from top to bottom, ionization energy decreases.
- More energy is needed for each succeeding ionization.
- Significantly more energy is needed to break a full shell of electrons.
- Elements with low ionization energies are more easily oxidized.

Electron Affinity

- Moving from left to right across a period, electron affinity increases.
- Moving down a group from top to bottom, electron affinity generally decreases.

Electronegativity

- Moving from left to right across a period, electronegativity increases.
- Moving down a group from top to bottom, electronegativity decreases.

CHEMICAL BONDING

BONDS BETWEEN ATOMS

Ionic Attractions

- **Ionic attractions** occur between elements—usually a combination of metal and non-metal atoms—that have a difference in electronegativity greater than or equal to 1.7.

- In an ionic attraction, an electron leaves the less electronegative atom—creating a positive charge, and migrates to the more electronegative atom—creating a negative charge. The unlike charges that result create an attraction between the two atoms.

- Ionic attraction follows Coulomb's Law. The strength of attraction is directly proportional to the amount of charge and inversely proportional to the square of the distance between the two charges.

- Ionic attractions satisfy the valence shells of the elements involved in the attraction. The metal loses electrons to have a filled shell. The non-metal gains electrons to have a filled shell.

Covalent Bonds

- **Covalent bonds** occur between elements—usually non-metals—that have a difference in electronegativity between 0 and 1.7.

- In covalent bonds, electrons are shared between two atoms. The attraction is created from the attraction between the opposite magnetic fields created by the two electrons in the bond. The number of covalent bonds that a non-metal may form equals eight minus the group number of the element.

- **Double and triple bonds.** Electrons in a covalent bond satisfy the valence shell octets of both elements in the bond. The first covalent bond between two non-metals is a **sigma bond** (σ), where the electrons are paired along the axis between the two atoms. Any additional covalent bonds between non-metals are **pi bonds** (π), where the electrons are paired through the sideways overlap of *p*-orbitals above and below the inter-nuclear axis. For example, a double bond consists of one sigma and one pi bond. Sigma bonds are much stronger than pi bonds, and are therefore more difficult to break.

- **Nonpolar covalent bonds** form when the difference in electronegativity of the two atoms is negligible (< 0.4). Electrons in the bond are shared equally. Typical examples of nonpolar bonds are those between diatomic molecules of the same element, such as F_2 or Cl_2.

- **Polar covalent bonds** form when the difference in electronegativity of the two atoms is between 0.4 and 1.7. In a polar covalent bond, the element with the greater electronegativity takes on a slightly negative charge (ς^-), and the element with the lesser electronegativity takes on a slightly positive charge (ς^+). In effect, the electrons in the bond are closer to the element with greater electronegativity.

Example:

$$\varsigma^+ \overset{(2.2)}{H} — \overset{(3.4)}{O} \varsigma^-$$

- A **dipole moment** exists in a polar covalent bond that points to the more electronegative atom. Dipole moments add together like vectors to create a total dipole moment on the molecule, as described in this chapter under "Structure and Physical Properties."

Example:

$$\varsigma^+ \overset{(2.2)}{H} — \overset{(3.4)}{O} \varsigma^-$$

Network Covalent

- Network covalent crystals are groups of non-metal atoms held together by covalent bonds. A commonly-cited example is a diamond, which is a network covalent crystal of carbon atoms.

- Network covalent crystals tend to have very high melting and boiling points, and a very high amount of energy is needed to break apart the crystal.

Metallic Attraction

- Metals are the elements on the periodic table to the left of the stairs.

- Outer electrons of metals are delocalized (shared between multiple bonds) and may be associated with one atom or another.

- Delocalized electrons give metals many of their physical properties: luster (because easily excited electrons give off photons as they return to the ground state), and electrical and thermal conductivity (because electrons move easily from atom to atom).

- Metallic bonding is created because all the metallic atoms share their outer electrons in a manner that are thought of as an "electron sea." The electrons, which are delocalized, move freely between the outer energy levels of the different atoms. This delocalization of electrons promotes electrical and thermal conductivity. Metals also demonstrate the qualities of malleability (can be shaped by hammering) and ductility (can be drawn into wire).

MOLECULAR MODELS

Covalent bonding can be modeled in a number of different ways, with each type of model having its own strengths and weaknesses.

Lewis Structures

- **Lewis structures** model the valence electrons that are involved in covalent bonding.

- The group number on the period table gives the number of valence electrons available for bonding. Normally, the "octet rule" must be satisfied to combine other atoms with a central atom—that is, the combination of the central atom's valence electrons and the electrons from other atoms bonded to the central atom should equal eight to create a stable Lewis structure.

- **Resonance structures** are an attempt to model delocalized electrons. For example, to allow the octet of sulfur in the sulfur dioxide molecule to be satisfied, it would show a double bond to one oxygen and not the other. However, the bonds to both oxygen atoms are equal, and demonstrate a bond length and strength somewhere between a single and a double bond. Therefore, the correct structure would be something between the two resonance structures below:

$$:S = O \quad \leftrightarrow \quad :S - O$$
$$\mid \qquad\qquad \parallel$$
$$O \qquad\qquad O$$

- Lewis structures are limited because they do not adequately depict delocalized electrons (except through resonance structures), show the difference between sigma and pi bonds, or explain the concept of expanded octets.

Hybridization, VSEPR, and Bond Geometry

- **Hybridization** refers to the process by which electrons mix traits of different atomic orbitals to create bonding orbitals. The hybridization model helps explain the difference in bond strength between sigma and pi bonds and correctly predicts the geometry of molecules. The following table details the relationship between the number of electrons around a central atom, its hybridization, and the geometry of the molecule.

- **VSEPR,** or *valence shell electron-pair repulsion*, dictates that electron pairs will repel each another, thus creating a molecular geometry in which each electron pair is as far as possible from every other electron pair. Unbonded pairs are slightly more repellant than other pairs. VSEPR theory gives a good explanation for the geometry found in the table below.

- **Expanded octets** are created when highly electronegative atoms bond to large central atoms and there is space to allow either five or six electron pairs around the central atom. Expanded octets require d-electrons to participate in hybridization (see table below).

Table — Summary of Hybridization

Number of Bonds	Number of Unused e pairs	Type of Hybrid Orbital	Angle between Bonded Atoms	Geometry	Example
2	0	sp	180°	Linear	BeF_2
3	0	sp^2	120°	Trigonal planar	BF_3
4	0	sp^3	109.5°	Tetrahedral	CH_4
3	1	sp^3	90° to 109.5°	Pyramidal	NH_3
2	2	sp^3	90° to 109.5°	Angular	H_2O
6	0	sp^3d^2	90°	octahedral	SF_6

Isomers

- **Isomers** are molecules that have the same formula but different structure—or arrangement of the atoms. Two isomers will have different physical and chemical properties, which will depend on how the atoms are arranged and the intermolecular forces that are created as a result.

  Example: Below are the two structural isomers of butane, each with the same formula, but different structure.

n-butane: $CH_3 — CH_2 — CH_2 — CH_3$

isobutane: $CH_3 — \overset{\displaystyle CH_3}{\underset{\displaystyle CH_3}{\overset{|}{\underset{|}{CH}}}} — CH_3$

ATTRACTIONS BETWEEN MOLECULES

Van der Waals Forces

- **Van der Waals forces** are sometimes referred to as dispersion forces or "instantaneous dipoles." These forces are created by the chance movement of electrons in a system of atoms bonded together.

- The strength of attraction is proportional to the number of electrons in the molecule. These forces are weak relative to polar intermolecular forces, and they become apparent only if there are many electrons, or if the molecules come very close together.

Polarity

- A **polar intermolecular attraction** can exist between two polar molecules. The slightly positive end of one molecule forms an electrostatic attraction to the slightly negative end of another molecule.

- A **hydrogen bond** occurs when a hydrogen atom is involved with a polar intermolecular attraction to a more electronegative atom.

- Polar intermolecular attractions are much stronger than van der Waals forces.

STRUCTURE AND PHYSICAL PROPERTIES

- **Physical properties are related to the forces between atoms and molecules in a crystal.** Hardness, melting point, and boiling point are all measures of the strength of interatomic or intermolecular attraction holding the crystal together. Stronger attraction results in lower vapor pressures.

- **Network covalent crystals** have the strongest attraction holding atoms together. The have the highest melting points and are the hardest of all crystals.

- **Ionic crystals** have strong electrostatic forces holding atoms together. They tend to have high melting and boiling points, and are poor conductors in the solid phase (because the electrons are in fixed positions). However, when ionic crystals are in the aqueous phase, they are good conductors because the charge is mobile.

- **Nonpolar molecules**, held together predominantly with van der Waals forces, are soft crystals, are easily deformed, and vaporize easily. They have much lower melting and boiling points than polar compounds of similar molar mass. Nonpolar molecules are poor conductors and tend to be more volatile because of their high vapor pressure.

- **Polar molecules** have intermolecular attractions that are weaker than ionic forces but are much stronger than nonpolar dispersion forces. The melting and boiling points of polar compounds depend on the strength of the dipole moment of the two compounds. The stronger the cumulative dipole moment, the greater the intermolecular attraction and the higher the melting and boiling points.

- **Hydrogen bonds** are the strongest polar intermolecular attraction because the hydrogen atom has only one electron. When that electron leaves, the exposed proton of the hydrogen nucleus is strongly attracted to the slightly negative charge of the other molecule. Consequently, polar molecules with hydrogen bonds have higher melting and boiling points than polar molecules of similar molar mass that do not have hydrogen bonds.

- **Summary of strength of attraction:**

  Covalent > ionic > metallic > polar intermolecular > nonpolar intermolecular

  Thus, if a covalent attraction needs to be broken to melt a solid, it will take more energy than if an ionic attraction needs to be broken, which will take more energy than a metallic attraction, and so on.

NUCLEAR CHEMISTRY

NUCLEAR REACTIONS AND EQUATIONS

Alpha Decay

- The nucleus emits a package of two protons and two neutrons, called an alpha particle (α), which is equivalent to the nucleus of a helium atom. This usually occurs with elements that have a mass number greater than 60.

- Alpha decay causes the atom's atomic mass to decrease by four units and the atomic number by two units.

- Example:

$$^{238}_{92}U \rightarrow \alpha \text{ particle } (^4_2He) + {}^{234}_{90}Th$$

Beta Decay

- The nucleus emits a beta particle (β^-) that degrades into an electron as it passes out of the atom. This usually occurs with elements that have a mass number greater than their atomic weight.

- Beta decay causes the mass number to remain the same but increases the atomic number by one. Beta decay converts a neutron into a proton.

- Example:

$$^{234}_{90}Th \rightarrow \beta^- \text{ particle } (^0_{-1}e) + {}^{234}_{91}Pa$$

Positron Decay

- The nucleus emits a particle that degrades into a positron as it passes out of the atom. This usually occurs with elements that have a mass number smaller than their atomic weight.

- Positron decay causes the mass number to remain the same but decreases the atomic number by one. Positron decay converts a proton into a neutron and a positron.

- Example:

$$^{13}_{7}N \rightarrow {}^0_{+1}e + {}^{13}_6C$$

Gamma Radiation

- Gamma rays (γ) are high-frequency, high-energy, electromagnetic radiation that are usually given off in combination with alpha and beta decay.

- Gamma decay can occur when a nucleus undergoes a transformation from a higher-energy state to a lower-energy state. The resulting atom may or may not be radioactive.

- Gamma rays are photons, which have neither mass nor charge.

RATE OF DECAY; HALF-LIFE

- Half-life is the time it takes for 50 percent of an isotope to decay.

- Nuclear decay represents a "first-order" reaction in that it depends on the amount of material and the rate constant.

- $$T_{1/2} = \frac{0.693}{k}$$

  where k is the decay constant.

- Example:

  Strontium-85 has a half-life of 65.2 days. How long will it take for 20 grams of strontium-85 to decay into five grams of strontium-85?

  Solution:

  It takes two half-lives to decrease the amount of strontium-85 from 20 grams to 5 grams.

  2×65.2 days $= 130.4$ days.

CHAPTER 3

The States of Matter

CHAPTER 3

THE STATES OF MATTER

GASES

Ideal Gas Laws

- **Ideal gases** are gases that behave according to an approximation that includes the following assumptions:

    1. The volume of the gas molecule is negligible compared to the space between the molecules.

    2. There is negligible intermolecular attraction between gas molecules.

- The ideal gas approximation is most accurate for gases at low pressure and high temperature.

- Under ideal conditions, the following laws hold true:

Boyle's Law

- **Boyle's law** states that the volume of a gas is inversely proportional to pressure, when temperature is constant.

- This can be summarized by the following expression:

$$P_1 V_1 = P_2 V_2$$

Example: A 4.0-liter elastic weather balloon travels from sea level, at 1.0 atm pressure, to a higher altitude, where the pressure is 0.20 atm. What is the new volume of the balloon?

Solution:

$$V_2 = \frac{P_1 V_1}{P_2} = \frac{1.0 \text{ atm} \times 4.0 \text{L}}{0.20 \text{ atm}} = 20.0 \text{ L}$$

Charles's Law

- **Charles's law** states that the volume of a given amount of gas is directly proportional to temperature, when pressure is constant.

- This can be summarized by the following expression:

$$V_1 T_2 = V_2 T_1$$

Example:

A gas occupies 2.0 L at 300 K. What is the volume of the gas at 200 K, assuming that the pressure is constant?

Solution:

$$V_2 = \frac{V_1 T_2}{T_1} = \frac{2.0 \text{ L} \times 200 \text{ K}}{300 \text{ K}} = 1.3 \text{ L}$$

Laws of Gay-Lussac

- The **law of Gay-Lussac** states that at constant volume, the pressure exerted by a given mass of gas varies directly with the absolute temperature.

- This can be summarized by the following expression:

$$P_1 T_2 = P_2 T_1$$

Example:

A gas in a rigid container exerts 6.0 atm at 300 K. What is the pressure that the gas exerts at 500 K?

Solution:

$$P_2 = \frac{P_1 T_2}{T_1} = \frac{6.0 \text{ atm} \times 500 \text{ K}}{300 \text{ K}} = 10 \text{ atm}$$

- Gay-Lussac's **law of combining volumes** states that when reactions take place in a gaseous state at constant temperature and pressure, the volume of reactants and products can be expressed as the ratios given by the stoichiometric coefficients in the balanced reaction.

Dalton's Law

- **Dalton's law** states that the total pressure exerted by a mixture of gases is equal to the sum of the partial pressures of the gases in the mixture.

$$P_{total} = P_1 + P_2 + P_3 + ...$$

- Dalton's law comes into play when a gas is collected over water, where the total pressure measured is equal to the pressure exerted by the collected gas plus the water vapor pressure at the temperature of the system.

Example:

A sample of methane gas is collected over water at an ambient pressure of 0.972 atm. The vapor pressure of water at this temperature is 0.025 atm. What is the pressure exerted by the methane?

Solution:

$P_{total} = P_{water} + P_{methane}$

$P_{methane} = P_{total} - P_{water} = 0.972 \text{ atm} - 0.025 \text{ atm} = 0.947 \text{ atm}$

- Partial pressures of individual gases in a gas mixture are proportional to the mole fraction of the gas in the mixture, which can be summarized by the following expression:

$$P_{gas\ a}\ n_{total} = P_{total}\ n_{gas\ a}$$

Example:

A rigid container with a combination of nitrogen and oxygen gas is at a pressure of 2.4 atm. If the mole fraction of nitrogen gas is 0.16, what is the partial pressure exerted by the nitrogen gas?

Solution:

$$P_{nitrogen} = \frac{P_{total}\ n_{nitrogen}}{n_{total}} = \frac{2.4 \text{ atm} \times 0.16 \text{ moles nitrogen}}{1.0 \text{ total moles}} = 0.38 \text{ atm}$$

Avogadro's Law

- **Avogadro's law** states that under conditions of constant temperature and pressure, the volume of a gas is proportional to the number of moles of gas present. This law can be summarized by the expression:

$$V_1\ n_2 = V_2\ n_1$$

Example:

Suppose you were given 8.00 moles of a gas occupying a volume of 4.00 L at constant pressure and temperature. What volume of gas would 16.0 moles occupy at the same temperature and pressure?

$V_1 = 4.00$ L $n_1 = 8.00$ mol

$V_2 = ?$ $n_2 = 16.0$ mol

Solution:

$$V_2 = \frac{V_1 n_2}{n_1}$$

$$V_2 = \frac{(4.00 \text{ L})(16.0 \text{ mol})}{8.00 \text{ mol}} = 8.00 \text{ L}$$

Combined Ideal Gas Law

- All of the laws mentioned above can be combined to show that the pressure multiplied by volume, divided by the number of gas moles and the temperature, equal a constant ratio, which is usually expressed by the **ideal gas law** in the following manner.

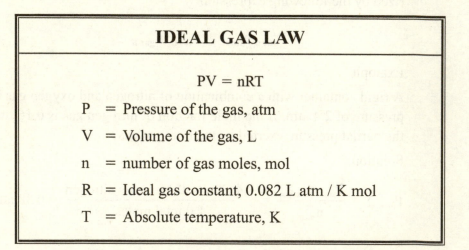

IDEAL GAS LAW
PV = nRT
P = Pressure of the gas, atm
V = Volume of the gas, L
n = number of gas moles, mol
R = Ideal gas constant, 0.082 L atm / K mol
T = Absolute temperature, K

Example:

What is the pressure exerted by 3.0 moles of gas at 200 K in a 2.0-liter container?

Solution:

$$P = \frac{nRT}{V} = \frac{(3.0 \text{ mole} \times 0.082 \text{ L atm} \times 200 \text{ K})}{(2.0 \text{ L}) \qquad \text{K mol}} = 24.6 \text{ atm}$$

- **Standard temperature and pressure (STP)** is 273 K and 1.0 atm. (This is not to be confused with the temperature of a system under thermodynamically standard conditions, which is 298 K and 1.0 atm.)

- 1.0 mole of gas at STP occupies a volume of 22.4 L, regardless of the identity of the gas.

- The ideal gas law can be rewritten in terms of the molar mass of a gas and its density using the following equation:

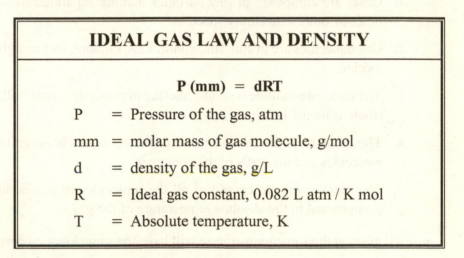

IDEAL GAS LAW AND DENSITY

P (mm) = dRT

P	= Pressure of the gas, atm
mm	= molar mass of gas molecule, g/mol
d	= density of the gas, g/L
R	= Ideal gas constant, 0.082 L atm / K mol
T	= Absolute temperature, K

Example:

A gas sample with a density of 1.67 g/L exerts a pressure of 2.0 atm at a temperature of 299 K. What is the molar mass of the gas?

Solution:

$$\text{Molar mass} = \frac{dRT}{P} = \frac{(1.67 \textbf{ g}) (0.082 \text{ L atm}) (299 \text{ K})}{\cancel{\text{L}} \qquad \cancel{\text{K}} \textbf{ mol} \, 2.0 \, \cancel{\text{atm}}} = 20.5 \text{ g/mol}$$

Example:

Compare the density of hydrogen gas with that of water vapor, at constant temperature and pressure.

Solution:

$$(\text{mm})_{\text{hydrogen}} \times d_{\text{water}} = (\text{mm})_{\text{water}} \times d_{\text{hydrogen}}$$

$$\frac{d_{water}}{d_{hydrogen}} = \frac{(mm)_{water}}{(mm)_{hydrogen}} = \frac{18}{2} = 9$$

Therefore, water vapor is nine times more dense than hydrogen gas at any given temperature and pressure.

KINETIC THEORY

Kinetic Molecular Theory

- Kinetic molecular theory states the following:

1. Gases are composed of tiny particles that are separated from each other by otherwise empty space.

2. Gas molecules are in constant, continuous, random, and straight-line motion.

3. The molecules collide with one another in perfectly elastic collisions (there is no net loss in energy).

4. The pressure of the gas is a result of the collisions between the gas molecules and the walls of the container.

5. The average kinetic energy of all the molecules of gas is directly proportional to the absolute temperature of the gas.

- All gases at the same temperature will have the same kinetic energy.

Graham's Law

- Graham's law is a result of the kinetic theory of matter, since two different gases at the same temperature and pressure, gas a and gas b, will both have the same kinetic energy.

$$\frac{1}{2} (m_a v_a^2) = \frac{1}{2} (m_b v_b^2)$$

where m_a = mass of gas a, v_a = velocity of gas a, m_b = mass of gas b, and v_b = velocity of gas b.

- **Effusion** is the process by which a gas escapes from one chamber to another by moving through a small hole. The rate of effusion is proportional to the velocity of the gas.

- **Graham's law of effusion** follows the format of the equation for kinetic energy. The law tells us that gas molecules of smaller molar mass move faster than gas molecules of larger molar mass, and that the faster moving molecules undergo effusion faster.

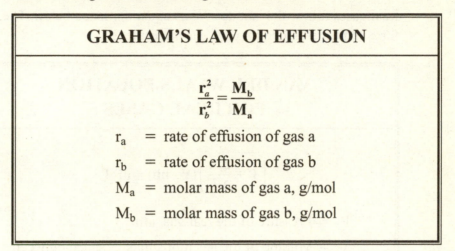

GRAHAM'S LAW OF EFFUSION

$$\frac{r_a^2}{r_b^2} = \frac{M_b}{M_a}$$

r_a = rate of effusion of gas a

r_b = rate of effusion of gas b

M_a = molar mass of gas a, g/mol

M_b = molar mass of gas b, g/mol

Example:

A mixture of helium and carbon dioxide form a mixture in a rigid container. A small leak is created in the container; how much faster will the helium exit the container than the carbon dioxide?

Solution:

$$\frac{M_{\text{carbon dioxide}}}{M_{\text{helium}}} = \frac{r_{helium}^2}{r_{carbon}^2} = \frac{44}{4} = 11$$

Therefore, helium leaves the container $(11)^{\frac{1}{2}}$ or 3.3 times faster than carbon dioxide.

DEVIATIONS FROM IDEAL GAS LAWS

- Real (non-ideal) gases occur when the volume of the molecule is significant relative to the space between the molecules, or when there is significant intermolecular attraction between molecules.

- If the volume of the space occupied by the molecules becomes significant, the measured volume of the gas will be larger than the volume calculated using the ideal gas law.

- If the intermolecular attraction between gas molecules becomes significant, the measured pressure of the gas will be less than the pressure calculated using the ideal gas law.

- Most gases are ideal under normal temperatures and pressures. However, under high pressures or low temperatures (when gas molecules come closer together), gases will be more likely to demonstrate real behavior.

- Real gases follow the van der Waals equation, listed below. The values for *a* and *b* are unique for each gas molecule.

VAN DER WAALS EQUATION FOR REAL GASES

$$\left(P + \frac{n^2 a}{V^2}\right)(V - nb) = nRT$$

P = Pressure of the real gas, atm

V = Volume of the gas molecules + space between molecules, L

n = number of gas moles, mol

R = Ideal gas constant, 0.082 L atm / K mol

T = Absolute temperature, K

a = intermolecular attraction constant for the gas

b = space occupied by one mole of gas molecules

LIQUIDS AND SOLIDS

PHASE DIAGRAMS

- A **phase diagram** shows the state of a substance (solid, liquid, or gas) at any given temperature and pressure.

- The lines on the diagram indicate those places where a substance exists simultaneously in both phases.

- The **critical point** is the temperature and pressure above which a substance must exist as a gas.

- The **triple point** is the temperature and pressure at which the substance may exist in all three phases: solid, liquid, and gas.

- The **vapor pressure curve** defines the boundary between the liquid and gas phases on the phase diagram. The vapor-pressure curve determines the partial pressure of gas that can be in the vapor phase at any given temperature.

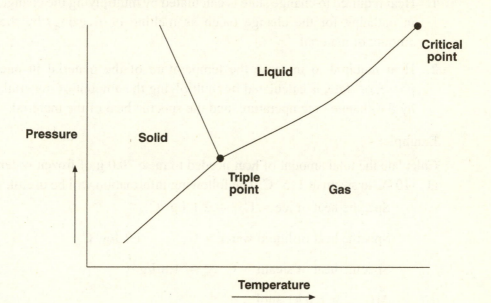

PHASE CHANGES

- Phase changes are caused by changes in pressure or temperature.
- Phase change involves a change in enthalpy (ΔH) that is unique for each substance and each change.

Change of phase	Name of change	ΔH for change
Solid to liquid	Melting (fusion)	$\Delta H > 0$
Liquid to solid	Freezing	$\Delta H < 0$
Liquid to gas	Vaporization	$\Delta H > 0$
Gas to liquid	Condensation	$\Delta H < 0$
Solid to gas	Sublimation	$\Delta H > 0$
Gas to solid	Deposition	$\Delta H < 0$

- A **heating curve** is a temperature-versus-time graph when heat is added at a constant rate, or a temperature-versus-heat-added graph. A heating curve in reverse is called a **cooling curve**. Two procedures are used to calculate the heat added or subtracted in a heating or cooling curve:

    1. Heat required to change state is calculated by multiplying the change in enthalpy for the change (such as melting or freezing) by the amount of material.

    2. Heat required to increase the temperature of the material in one phase, or state, is calculated by multiplying the amount of material, by the change in temperature, and the specific heat of the material.

Example:

Calculate the total amount of heat needed to raise 20.0 g of frozen water at -10 °C to steam at 115 °C. The following information will be useful:

$$\text{Specific heat of ice} = C_{ice} = 2.1 \text{ J/g °C}$$

$$\text{Specific heat of liquid water} = C_{water} = 4.2 \text{ J/g °C}$$

$$\text{Specific heat of steam} = C_{steam} = 1.8 \text{ J/g °C}$$

$$\Delta H_{fusion} = 6.0 \text{ kJ /mol}$$

$$\Delta H_{vaporization} = 40.7 \text{ kJ/mol}$$

Solution:

The answer equals the sum of the heats of the following processes.

Heat needed to raise ice to melting point =

$$20 \text{ g} \times \frac{2.1 \text{ J}}{\text{g °C}} \times 10 \text{ °C} = 420 \text{ J}$$

Heat needed to melt ice =

$$20.0 \text{ g} \times \frac{1.0 \text{ mol}}{18.0 \text{ g}} \times \frac{6.0 \text{ kJ}}{\text{mol}} = 6.66 \text{ kJ}$$

Heat needed to raise water to boiling point =

$$20 \text{ g} \times \frac{4.2 \text{ J}}{\text{g °C}} \times 100° \text{ C} = 8400 \text{ J}$$

Heat needed to boil water =

$$20 \text{ g} \times \frac{1.0 \text{ mol}}{18.0 \text{ g}} \times \frac{40.7 \text{ kJ}}{\text{mol}} = 45.2 \text{ kJ}$$

Heat needed to raise steam to final temperature =

$$20 \text{ g} \times \frac{1.8 \text{ J}}{\text{g }^{\circ}\text{C}} \times 15^{\circ} \text{ C} = 540 \text{ J}$$

Total heat needed =

0.420 kJ + 6.66 kJ + 8.40 kJ + 45.2 kJ + 0.54 kJ = <u>61.22 kJ</u>

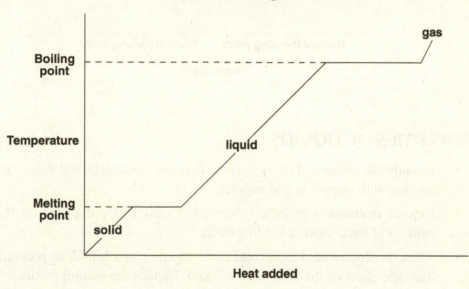

Typical Heating Curve

- The **normal melting point** is the temperature that corresponds to the solid-liquid equilibrium at 1.0 atm pressure.
- The **normal boiling point** is the temperature that corresponds to the liquid-gas equilibrium at 1.0 atm pressure.

Boiling and Freezing Points on a Phase Diagram

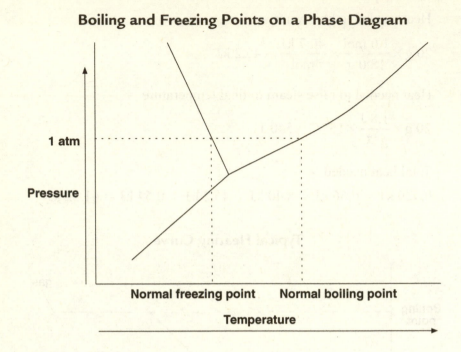

PROPERTIES OF LIQUIDS

- Liquids are composed of molecules that are constantly and randomly moving with respect to one another.

- Liquids maintain a definite volume, but their shape depends on the contour of the container holding them.

- Attractive forces hold molecules close together in a liquid, so pressure has little effect on the volume of a liquid. Liquids are incompressible.

- Changes in temperature cause only small changes in volume.

- Liquids diffuse more slowly than gases. Diffusion increases as temperature increases.

- Surface tension is the inward force of a liquid toward itself. Surface tension decreases as temperature increases.

PROPERTIES OF SOLIDS

- Solids retain their shape and volume when transferred from one container to another.

- Solids are virtually incompressible.

- The attractive forces between atoms, molecules, or ions in a solid are relatively strong. The particles are held in a fixed position relative to one another.

- **Crystalline solids** are composed of structural units bounded by a specific geometric pattern and have sharp melting points. Table salt (NaCl) is a crystalline solid. The following are different configurations of atoms in crystals:

  1. A **unit cell** is the smallest repeating unit in a crystalline solid.

  2. **Simple cubic** unit cells have one atom at each of the corners of the cube. One-eighth of each of the eight atoms is inside the cube, so a simple cubic contains a total of one atom per unit cell.

  3. **Face-centered crystal** is a simple cubic unit cell, with one additional atom shared between two unit cells on each face of the cube. There are a total of four atoms per unit cell.

  4. **Body-centered crystal** is a simple cubic unit cell, with one additional atom in the center of the cube. There is a total of two atoms per unit cell.

- **Amorphous solids** do not display a specific geometry and do not have a sharp melting point. Glass is an amorphous solid.

SOLUTIONS

SOLUBILITY

Solution Process

- **Solvation** is the interaction of solvent molecules with solute molecules (or ions) to form loosely bonded aggregates of both solvent and solute. The attraction between solute and solvent must be greater than the force that holds the solute together in order for the solute to dissolve.

- **Hydration** describes the solvation process when water is the solvent.

- **Miscible** solutions occur when one substance is soluble in all proportions with another substance.

Factors That Affect Solubility

Concentration

- A **saturated solution** occurs when a solid solute is in equilibrium with dissolved solute.

- The **solubility** of a solute is the molar concentration of dissolved solute at saturation.

- **Supersaturated solutions** contain more solute than required for saturation.

Temperature

- The solubility of solids in liquids increases with increasing temperature when the $\Delta H_{solvation}$ is endothermic.

- The solubility of solids in liquids decreases with increasing temperature when the $\Delta H_{solvation}$ is exothermic.

- For gases in liquids, the solubility usually decreases with increasing temperature.

Pressure: Henry's Law

- While pressure does not appreciably affect the solubility of liquids or solids in liquids, Henry's law determines how much the solubility of gases in liquids increases when the partial pressure of the same gas above the liquid increases.

- Henry's law states that the amount of gas that can dissolve in a liquid is directly proportional to the partial pressure of the gas above the liquid.

- Henry's law is most accurately obeyed for gases that do not dissociate or react with the liquid.

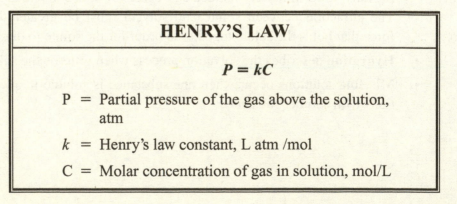

HENRY'S LAW

$$P = kC$$

P = Partial pressure of the gas above the solution, atm

k = Henry's law constant, L atm /mol

C = Molar concentration of gas in solution, mol/L

Example:

The partial pressure of carbon dioxide at sea level is 0.0004 atm. The Henry's law constant for carbon dioxide is 32.0 L atm/mol at 25 °C. What is the molar concentration of carbon dioxide in a glass of water at sea level?

Solution:

$$C = \frac{P}{k} = \frac{0.0004 \text{ atm mol}}{32.0 \text{ L atm}} = 1.25 \times 10^{-5} \text{ M}$$

SOLUTION CONCENTRATIONS

Molarity

$$\text{Molarity (M)} = \frac{\textbf{Moles Solute}}{\textbf{Liters Solution}}$$

Example:

What is the molarity of a solution when 10.0 g of HCl is completely dissolved in dissolved water to make 500 mL of solution? What is the pH of the solution?

Solution:

$$0.55 \text{ M HCl} = 10.0 \text{ g HCl} \times \frac{1 \text{ mole}}{36.45 \text{ g}} \times \frac{1}{0.5 \text{ L}}$$

$$\text{pH } 0.26 = -\log [0.55]$$

pH

- pH is defined as the negative log of the molar hydrogen ion concentration. This means that the pH is the negative of the exponent to which ten is raised for the molar hydrogen ion, or hydronium ion concentration.

$$\textbf{pH} = -\textbf{log [H}^+\textbf{]}$$

Examples:

[H$^+$]	pH
1×10^{-1}	1
1×10^{-2}	2
1×10^{-8}	8
1×10^{-10}	10
1×10^{-14}	14

Molality

$$\text{Molality} = \frac{\text{Moles Solute}}{\text{Kilograms Solvent}}$$

Example:

What is the molality of the solution that is created when 10.0 g of NaCl is added to 800 g of water?

Solution:

$$0.214 \text{ molal} = 10.0 \text{ g solute} \times \frac{1 \text{ mole solute}}{58.45 \text{ g NaCl}} \times \frac{1}{0.8 \text{ kg solvent}}$$

Mole Fraction

$$\text{mole fraction} = \frac{\text{moles solute}}{\text{total solution moles}}$$

Example:

1.00 g ethanol is mixed with 100.0 g pure water. What is the mole fraction of the ethanol?

Solution:

$$0.0217 \text{ moles ethanol} = 1.000 \text{ g ethanol} \times \frac{1.0 \text{ mol ethanol}}{46.07 \text{ g ethanol}}$$

$$5.555 \text{ moles water} = 100.0 \text{ g water} \times \frac{1.0 \text{ mole water}}{18.00 \text{ g water}}$$

$$\text{Ethanol mole fraction} = \frac{\text{moles ethanol}}{\text{mol water} + \text{mol ethanol}} =$$

$$\frac{0.0217}{5.5772} = 0.00389$$

COLLIGATIVE PROPERTIES

Raoult's Law

- **Raoult's law** describes how the vapor pressure curve of a solvent is depressed when a solute dissolves in the solvent.

- A depressed vapor pressure curve also causes other colligative properties to change, such as boiling and freezing points and osmotic pressure.

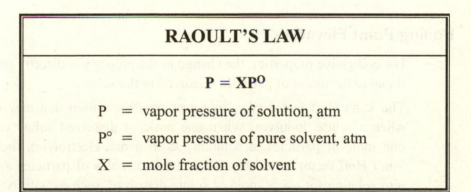

> # RAOULT'S LAW
>
> $$P = XP^O$$
>
> P = vapor pressure of solution, atm
>
> P° = vapor pressure of pure solvent, atm
>
> X = mole fraction of solvent

Example:

What is the vapor pressure of an aqueous solution that is created when 400.0 g of glucose are dissolved in 500.0 g of water at 25 °C? (The vapor pressure of pure water at 25 °C is 0.031 atm.) The molecular weight of glucose is 180 g/mol.

Solution:

$$\text{Moles glucose} = \frac{400.0 \text{ g glucose} \times \text{mol}}{180 \text{ g glucose}} = 2.22 \text{ moles glucose}$$

$$\text{Moles water} = \frac{500.0 \text{ g water} \times \text{mol}}{18 \text{ g water}} = 27.77 \text{ moles water}$$

Moles fraction of water $= \dfrac{\text{moles water}}{\text{Total moles}} = \dfrac{27.77}{29.99} = 0.93$

Vapor pressure of solution $= XP^O = (0.93)(0.031 \text{ atm}) = \underline{0.029 \text{ atm}}$

Example:

20.0 g of an unknown is dissolved in 125 g of pure water, and shows a vapor pressure of 0.030 atm at 25 °C. What is the molar mass of the unknown?

Solution:

mole fraction of water $= \dfrac{P_{solution}}{P_{pure\ water}} = \dfrac{\text{moles water}}{\text{mole water + mole unknown}}$

moles unknown = 0.23 moles

unknown molar mass $= \dfrac{20.0 \text{ g}}{0.23 \text{ mol}} = 86.96 \text{ g/mol}$

Boiling Point Elevation

- For colligative properties, the change in the property is directly proportional to the moles of particles dissolved in the solute.

- The Van't Hoff factor takes into account dissociation that may occur when a solute dissolves. When one mole of dissolved solute creates one mole of particles in solution (as in a non-electrolyte), then the Van't Hoff factor would be 1.0. When two moles of particles are put into solution for every mole of solute dissolved, such as with NaCl in water, then the Van't Hoff factor would be 2.0.

BOILING POINT ELEVATION

$$\Delta T = k_b m i$$

ΔT = Change in solvent boiling point, °C

k_b = molal boiling point constant of the solvent, °C kg/mol

m = molality of solute, mol/kg

i = Van't Hoff factor of solute

Example:

Compare the effects of 10.0 g sucrose (158.0 g/mol) with 10.0 of magnesium chloride (95.2 g/mol) in changing the boiling point of 100.0 g of water. (The molal boiling point constant of water is 0.51 °C kg/mol.)

Solution:

Note that the Van't Hoff factor for sucrose is 1.0, since it is a non-electrolyte. However, the Van't Hoff factor for magnesium chloride is 3.0, since it completely dissociates in water.

$$\Delta T \text{ due to sucrose} = \frac{0.51 \text{ °C kg}}{\text{mol}} \times \frac{10.0 \text{ g sucrose mol}}{158.0 \text{ g} \times 0.10 \text{ kg solvent}} \times$$

$$1.0 = 0.3 \text{ °C}$$

$$\Delta T \text{ due to MgCl}_2 = \frac{0.51 \text{ °C kg}}{\text{mol}} \times \frac{10.0 \text{ g MgCl}_2 \text{ mol}}{95.2 \text{ g} \times 0.10 \text{ kg solvent}} \times$$

$$3.0 = 1.6 \text{ °C}$$

The sucrose raises the boiling point by 0.3 °C, whereas the magnesium chloride raises the boiling point by 1.6 °C.

Freezing Point Depression

<table>
<tr><td colspan="2">FREEZING POINT DEPRESSION</td></tr>
<tr><td colspan="2">$\Delta T = k_f m i$</td></tr>
<tr><td>ΔT =</td><td>Change in solvent freezing point, °C</td></tr>
<tr><td>k_f =</td><td>molal freezing point constant of the solvent, °C kg/mol</td></tr>
<tr><td>m =</td><td>molality of solute, mol/kg</td></tr>
<tr><td>i =</td><td>Van't Hoff factor of solute</td></tr>
</table>

Example:

Enough magnesium chloride is placed on a cold wet road to create a 2.5 molal solution with the water on the road. At what temperature will ice form on the road? (The molal freezing point constant for water is 1.86°C kg/mol.)

Solution: The Van't Hoff factor for magnesium chloride is 3.0 since it completely dissolves in water; K_f = 1.86 °C kg/mol and m = 2.5 mol/kg.

$$\Delta T = k_f mi = \frac{1.86 \text{ °C kg}}{\text{mol}} \times \frac{2.5 \text{ mol}}{\text{kg}} \times 3.0 = 14 \text{ °C}$$

Since the freezing point of pure water is 0.0 °C, then the freezing point of the solution created by the magnesium chloride is −14 °C.

Osmotic Pressure

OSMOTIC PRESSURE

$$\pi V = nRTi$$

π = osmotic pressure, atm

V = volume of the solution, L

n = moles of solute, mol

R = ideal gas constant, 0.082 L atm / K mol

T = temperature, K

i = Van't Hoff factor of the solute

Example:

Exactly one gram (1.000 g) of an unknown protein was dissolved in enough pure water to make 500 mL of solution. The resulting osmotic pressure exerted by the solution was 0.002 atm at 298 K. What is the molar mass of the protein? (Assume that the protein is a non-electrolyte.)

Solution:

$$\text{Molar mass} = \frac{\text{mass solute} \times R\ T}{\pi\ V} =$$

$$\frac{1.000\ g \times 0.082\ L\ atm \times 298\ K}{0.002\ atm \times K\ mol \times 0.5\ L} = 24,436\ g/mol$$

NONIDEAL SOLUTIONS

- **Ideal solutions** follow Raoult's Law. The vapor pressure of a solution is directly proportional to the mole fraction of the solvent in the solution.

- **Nonideal solutions** experience a vapor pressure that is different from that predicted by Raoult's Law.

- **Negative deviations** from Raoult's Law occur when there is a stronger solute-solvent attraction, such as hydrogen bonding, that prevents solvent molecules from escaping into the vapor phase. Negative deviations also occur when the $\Delta H_{solution}$ is large and exothermic.

- **Positive deviations** from Raoult's Law occur when both the solute and solvent are very volatile, and also when $\Delta H_{solution}$ is large and endothermic.

CHAPTER 4

Types of Reactions

CHAPTER 4

TYPES OF REACTIONS

ACID-BASE REACTIONS

- **Arrhenius Theory** states that acids are substances that ionize in water to donate protons (H^+), and bases produce hydroxide ions (OH^-) when put into water.

  Example:

  Arrhenius acid: $HC_2H_3O_2 \rightarrow C_2H_3O_2^- + H^+$

  Arrhenius base: $NH_3 + H_2O \rightarrow NH_4^+ + OH^-$

- **Brønsted-Lowry Theory** states that acids donate protons—like an Arrhenius acid—and bases accept protons. Simply put, a proton moves from one compound to another. Each compound in a conjugate acid-base pair is different from each other by the existence of a proton.

  Example:

  $HC_2H_3O_2 + OH^- \rightarrow C_2H_3O_2^- + H_2O$

- **Lewis Theory** defines an acid as an electron-pair acceptor and a base as an electron-pair donor. The Lewis definition of acid-base reactions allows the inclusion of reactions that may not involve protons, such as the formation of coordination complexes.

  Example:

  $BF_3 + NH_3 \rightarrow BF_3NH_3$

- **Neutralization** is the process where an Arrhenius acid and base are combined to form a salt and water.

  Example:

  $HCl + NaOH \rightarrow H_2O + NaCl$

- **Amphoteric** compounds can act as either acids or bases.

  Example:

  $Al(OH)_3 + OH^- \rightarrow [Al(OH)_4]^-$

  $Al(OH)_3 + 3\,H^+ \rightarrow Al^{3+} + 3\,H_2O$

PRECIPITATION REACTIONS

- Precipitation reactions occur when soluble reactants are mixed together to form an insoluble product, according to the solubility rules below.

  Example:

  $Pb^{2+} (aq) + 2\,I^- (aq) \rightarrow PbI_2 (s)$

- Precipitation reactions are best written as **net ionic reactions**, where only the ions that combine to form the precipitate are shown. All other ions that remain dissolved in solution are **spectator ions**.

  Example:

  Full equation: $Pb(NO_3)_2 (aq) + 2\,KI (aq) \rightarrow PbI_2 (s) + 2\,KNO_3 (aq)$

  Net ionic reaction: $Pb^{2+} (aq) + 2\,I^- (aq) \rightarrow PbI_2 (s)$

Solubility Rules for Ions in Solution

- All compounds with IA metals and ammonium are **soluble**.
- All **nitrates are soluble**.
- All **chlorates and perchlorates are soluble**.
- All **acetates are soluble**.
- All **halides are soluble**, EXCEPT those that combine with Ag^+, Pb^{2+}, and Hg_2^{2+}.
- All **sulfates are soluble**, EXCEPT those that combine with Ag^+, Pb^{2+}, and Hg_2^{2+}, Ca^{2+}, Sr^{2+}, Ba^{2+}.
- All **hydroxides are insoluble**, EXCEPT those with IA metals, ammonium, Ca^{2+}, Sr^{2+}, Ba^{2+}.
- All **carbonates are insoluble**, EXCEPT those that contain IA metals and ammonium.
- All **phosphates are insoluble**, EXCEPT those that contain IA metals and ammonium.
- All **sulfites are insoluble**, EXCEPT those that contain IA metals and ammonium.
- All **chromates are insoluble**, EXCEPT those that contain IA metals and ammonium.
- All **sulfides are insoluble**, EXCEPT those that contain IA metals, IIA metals, and ammonium.

OXIDATION-REDUCTION REACTIONS

- **Oxidation** occurs when an atom increases control over an electron, or shows an increase in oxidation state. The oxidized species causes some other compound to be reduced, and is therefore called the **reducing agent**.

- **Reduction** occurs when an atom decreases control over an electron or shows a reduced oxidation state. The reduced species causes some other compound to be oxidized, and is therefore called the **oxidizing agent**.

- Oxidation-reduction reactions can be either spontaneous (and occur without an input of energy) or non-spontaneous (occur only with the input of energy).

- The number of electrons lost in oxidation always equals the number of electrons gained in reduction.

- When the oxidation and reduction half-reactions are physically separated, oxidation occurs at the **anode**, and reduction occurs at the **cathode**.

- Use "**Leo** the lion says **ger**" to help you remember—
  Loss of **e**lectrons, **o**xidation; **g**ain of **e**lectrons, **r**eduction.

Oxidation States

- **Oxidation state** is a term that is used interchangeably with the term **oxidation number**.

- Assigning oxidation numbers can help determine which compound is oxidized, and which compound is reduced. An increase in oxidation number during a reaction indicates oxidation; a decrease in oxidation number shows reduction.

- Oxidation numbers can help balance oxidation-reduction reactions (see the Stoichiometry chapter regarding balancing oxidation-reduction reactions.)

- Rules for assigning oxidation numbers:

  1. The total oxidation number of any element, ion, or compound equals the charge on that element, ion, or compound.

  2. Except in the case of hydrides, hydrogen has an oxidation state of $+1$ when combined with other atoms.

  3. Except in peroxides, oxygen has an oxidation state of -2 when combined with other atoms.

  4. IA, IIA, and VIIA elements tend to take the oxidation state of their most common ion when combined with other atoms.

Types of Oxidation-Reduction Reactions

- **Two uncombined elements come together**

  Example: $Mg + O_2 \rightarrow MgO$

- **Combustion with oxygen**

  Example: $C_4H_{10} + O_2 \rightarrow CO_2 + H_2O$

- **Decomposition of a single reactant**

  Example: $H_2O_2 \rightarrow O_2 + H_2O$

- **A solid transition metal is placed in a solution of metallic ions.** Use the chart of standard reduction potentials. The change with the highest reduction potential is reduced in charge. Voltaic or galvanic cells are an example of this type of reaction; see the next section.

  Example: $Cu + Ag^+ \rightarrow Cu^{++} + Ag$

- **Electrolytic cells**

  Example: $NaCl\ (l) \rightarrow Na\ (l) + Cl_2\ (g)$

- **A solid metal is added to an acid**

  The metal is oxidized and hydrogen gas is formed. Remember, water can also be an acid; check to be sure the dissociation of water on the reduction potential chart shows a higher tendency to be reduced, such as the case with calcium in water.

  Example: $Mg + H^+ \rightarrow Mg^{++} + H_2$

Voltaic (Galvanic) Cells

- **Voltaic, or galvanic cells** convert chemical energy into electrical energy by isolating the oxidation and reduction half-reactions.

- The **electromotive force** is the force with which the electrons flow through an external wire from the negative electrode to the positive electrode.

- **Oxidation occurs at the anode**, which is negatively charged for voltaic cells.

- **Reduction occurs at the cathode**, which is positively charged for voltaic cells.

- The greater the tendency for the two half-reactions to occur spontaneously, the greater the cell potential voltage, or E_{cell}, which is measured in volts (V).

- Cells are identified in shorthand by the oxidation half-reaction first. Example:

  The shorthand notation for the reaction, $Cu + Ag^+ \rightarrow Cu^{++} + Ag$, is $Cu / Cu^{++} // Ag^+/Ag$.

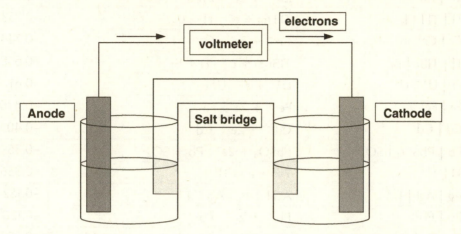

A voltaic cell separates the two half-reactions in a spontaneous oxidation-reduction reaction. Oxidation takes place at the anode; reduction takes place at the cathode. A salt bridge allows for a complete circuit.

Table — Standard Electrode Potentials in Aqueous Solutions at 25°C.

Electrode	Electrode Reaction	E°_{red}
Acid Solutions		
Li | Li$^+$	Li$^+$ + e$^-$ Li	−3.045
K | K$^+$	K$^+$ + e$^-$ K	−2.925
Ba | Ba^{2+}	Ba^{2+} + 2e$^-$ Ba	−2.906
Ca | Ca^{2+}	Ca^{2+} + 2e$^-$ Ca	−2.87
Na | Na$^+$	Na$^+$ + e$^-$ Na	−2.714
La | La^{3+}	La^{3+} + 3e$^-$ La	−2.52
Mg | Mg^{2+}	Mg^{2+} + 2e$^-$ Mg	−2.363
Th | Th^{4+}	Th^{4+} + 4e$^-$ Th	−1.90
U | U^{3+}	U^{3+} + 3e$^-$ U	−1.80
Al | Al^{3+}	Al^{3+} + 3e$^-$ Al	−1.66
Mn | Mn^{2+}	Mn^{2+} + 2e$^-$ Mn	−1.180

(Continued)

Table — Standard Electrode Potentials in Aqueous Solutions at 25°C. (*Continued*)

Electrode	Electrode Reaction	$E°_{red}$
$V \mid V^{2+}$	$V^{2+} + 2e^- \ V$	–1.18
$Zn \mid Zn^{2+}$	$Zn^{2+} + 2e^- \ Zn$	–0.763
$Tl \mid TlI \mid I^-$	$TlI(s) + e^- \ Tl + I^-$	–0.753
$Cr \mid Cr^{3+}$	$Cr^{3+} + 3e^- \ Cr$	–0.744
$Tl \mid TlBr \mid Br^-$	$TlBr(s) + e^- \ Tl + Br^-$	–0.658
$Pt \mid U^{3+}, U^{4+}$	$U^{4+} + e^- \ U^{3+}$	–0.61
$Fe \mid Fe^{2+}$	$Fe^{2+} + 2e^- \ Fe$	–0.440
$Cd \mid Cd^{2+}$	$Cd^{2+} + 2e^- \ Cd$	–0.403
$Pb \mid PbSO_4 \mid SO_4^{2-}$	$PbSO_4 + 2e^- \ Pb + SO_4^{2-}$	–0.359
$Tl \mid Tl^+$	$Tl^+ + e^- \ Tl$	–0.3363
$Ag \mid AgI \mid I^-$	$AGI + e^- \ Ag + I^-$	–0.152
$Pb \mid Pb^{2+}$	$Pb^{2+} + 2e^- \ Pb$	–0.126
$Pt \mid D_2 \mid D^+$	$2D^+ + 2e^- \ D_2$	–0.0034
$Pt \mid H_2 \mid H^+$	$2H^+ + 2e^- \ H_2$	–0.0000
$Ag \mid AgBr \mid Br^-$	$AgBr + e^- \ Ag + Br^-$	+0.071
$Ag \mid AgCl \mid Cl^-$	$AgCl + e^- \ Ag + Cl^-$	+0.2225
$Pt \mid Hg \mid Hg_2Cl_2 \mid Cl^-$	$Hg_2Cl_2 + 2e^- \ 2Cl^- + 2Hg(l)$	+0.2676
$Cu \mid Cu^{2+}$	$Cu^{2+} + 2e^- \ Cu$	+0.337
$Pt \mid I_2 \mid I^-$	$I_2^- + 2e^- \ 3I^-$	+0.536
$Pt \mid O_2 \mid H_2O_2$	$O_2 + 2H^+ 2e^- \ H_2O_2$	+0.682
$Pt \mid Fe^{2+}, Fe^{3+}$	$Fe^{3+} + e^- \ Fe^{2+}$	+0.771
$Ag \mid Ag^+$	$Ag^+ + e^- \ Ag$	+0.7991
$Au \mid AuCl_4^-, Cl^-$	$AuCl_4^- + 3e^- \ Au + 4Cl^-$	+1.00
$Pt \mid Br_2 \mid Br^-$	$Br_2 + 2e^- \ 2Br^-$	+1.065
$Pt \mid Tl^+, Tl^{3+}$	$Tl^{3+} + 2e^- \ Tl^+$	+1.25
$Pt \mid H^+, Cr_2O_7^{2-}, Cr^{3+}$	$Cr_2O_7^{2-} + 14H^+ 6e^- \ 2Cr^{3+} + 7H_2O$	+1.33
$Pt \mid Cl_2 \mid Cl^-$	$Cl_2 + 2e^- \ 2Cl^-$	+1.3595
$Pt \mid Ce^{4+}, Ce^{3+}$	$Ce^{4+} + e^- \ Ce^{3+}$	+1.45
$Au \mid Au^{3+}$	$Au^{3+} + 3e^- \ Au$	+1.50
$Pt \mid Mn^{2+}, MnO_4^-$	$MnO_4^- + 8H^+ + 5e^- \ Mn^2 + 4H_2O$	+1.51
$Au \mid Au^+$	$Au^+ + e^- \ Au$	+1.68

(*Continued*)

Table — Standard Electrode Potentials in Aqueous Solutions at 25°C. (*Continued*)

Electrode	Electrode Reaction	$E°_{red}$
$PbSO_4$ | PbO_2 | H_2SO_4	$PbO_2 + SO_4 + 4H^+ + 2e^-$ $PbSO_4 + 2H_2O$	+1.685
Pt | F_2, F^-	$F_2(g) + 2e^-$ $2F^-$	+2.87
Basic Solutions		
Pt | SO_3^{2-}, SO_4^{2-}	$SO_4^{2-} + H_2O + 2e^-$ $SO_3^{2-} + 2OH^-$	−0.93
Pt | H_2 | OH^-	$2H_2O + 2e^-$ $H_2 + 2OH^-$	−0.828
Ag | $Ag(NH_3)_2^+, NH_3(aq)$	$Ag(NH_3)_2 + e^-$ $Ag + 2NH_3$ (aq)	+0.373
Pt | O_2 | OH^-	$O_2 + 2H_2O + 4e^-$ $4OH^-$	+0.401
Pt | MnO_2 | MnO_4^-	$MnO_4^- + 2H_2O + 3e^-$ $MnO_2 + 4OH^-$	+0.588

Calculating Cell Voltage

Example:

Find the standard cell voltage for the reaction, $Cu + Ag^+ \rightarrow Cu^{++} + Ag$.

Solution:

1. Break the reaction into the oxidation and reduction half-reactions.

   Oxidation: $Cu \rightarrow Cu^{++} + 2e^-$

   Reduction: $e^- + Ag^+ \rightarrow Ag$

2. Use the chart of standard reduction potentials to find the reduction potential ($E°_{red}$) for each half-reaction. Since $E°_{ox} = - E°_{red}$, reverse the sign of the reduction potential for the oxidation half-reaction, and then add the two potentials together for the total cell potential under standard conditions. (Note that the oxidation or reduction potentials are not multiplied by the stoichiometric coefficient, as would be heats of reaction or entropies.)

Oxidation:	Cu	$\rightarrow Cu^{++} + 2e^-$	$E°_{ox}$	= −0.34 V
Reduction:	$e^- + Ag^+$	$\rightarrow Ag$	$E°_{red}$	= 0.80 V
Total reaction:	$Cu + Ag^+$	$\rightarrow Cu^{++} + Ag$	$E°_{total}$	= 0.46 V

Nernst Equation

- **The Nernst equation** is used to calculate the cell voltage under non-standard conditions.

THE NERNST EQUATION

$$E_{cell} = E°_{cell} - RT/nF \ln Q$$

$$E_{cell} = E°_{cell} - 0.06/n \ln Q$$

E_{cell} = cell voltage under non-standard conditions, Volts

$E°_{cell}$ = cell voltage under standard conditions, Volts

R = Ideal gas constant, 8.31 V C / mol K

T = absolute temperature, K

n = moles of electrons transferred in the reaction, mol

F = Faraday's constant, 96,486 C / mol

Q = reaction quotient for the reduction half-reaction

Example:

What is the total cell voltage for the reaction, $Cu + Ag^+ \rightarrow Cu^{++} + Ag$, when $[Ag^+] = 0.10$ M and $[Cu^{++}] = 2.0$ M? Use equilibrium principles to explain the difference between this cell potential and the cell potential at standard conditions.

Solution:

$E_{cell} = E°_{cell} - 0.06/n \ln Q = (0.46) - [0.06/2 \ln (2.0/0.10)] = 0.34$ V

The cell is non-standard in that the concentrations of the product are above 1.0 M and the concentrations of the reactant are below 1.0 M. The law of mass action suggests that the reaction will be less likely to go in a forward direction, and so will have less cell potential in the forward direction.

Electrolytic Cells

- **Electrolytic cells** use electrons from a power source to force an otherwise non-spontaneous oxidation-reduction reaction to occur.

- The cathode is negatively charged, and the anode is positively charged. Oxidation takes place at the anode, and reduction takes place at the cathode.

- It is important to have knowledge of two aspects of electrolytic cells: knowledge of the *type of reactions* that occur at the cathode and anode, and knowledge of *how much* reactant will be used up with a given amount of electrical charge.

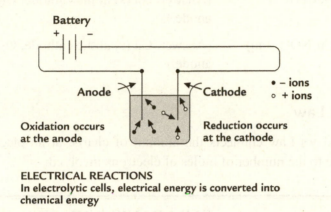

ELECTRICAL REACTIONS
In electrolytic cells, electrical energy is converted into chemical energy

- The following table summarizes the types of reactions that occur in an electrolytic cell. Examine the reactants that are available; the highest priority half-reaction will occur at each electrode. For example, if both water and chloride ions are available, the chloride ions will be oxidized and used up before water is oxidized.

Oxidation Reactions at the Anode	Reduction Reactions at the Cathode
First priority: halogen ions form gases.	First priority: transition metal ions form solid.
e.g.: Cl^2 (aq) $\rightarrow$ Cl_2 (g)	*e.g.:* Cu^+ (aq) $\rightarrow$ Cu (s)
Second priority: water ionizes to form gas.	Second priority: water ions form gases.
e.g.: H_2O (l) $\rightarrow$ $2H^+$ + O_2 (g)	*e.g.:* $2H_2O$ (l) $\rightarrow$ $2OH^-$ (aq) + H_2 (g)
	Third priority: molten metal ions form solid.
	e.g.: Na^+ (l) $\rightarrow$ Na (s)

Examples:

Determine what material will be produced at the cathode and anode when each of the following are placed in an electrolytic cell.

1. NaCl (aq)   (Answer: H_2 (g) at the cathode, Cl_2 (g) at the anode.)

2. NaCl (l)   (Answer: Na (s) at the cathode, Cl_2 (g) at the anode.)

3. $Cu(NO_3)_2$ (aq)   (Answer: Cu (s) at the cathode, O_2 (g) at the anode.)

Faraday's Law

• Faraday's Law connects the amount of charge at an electrolytic electrode to the number of moles of electrons involved.

FARADAY'S LAW
1.0 Faraday (F) = 1.0 mole electrons = 96,486 Coulombs

• Faraday's Law can be used along with the following definition of charge to solve a number of different types of questions.

CURRENT, CHARGE, AND TIME
$$I = \frac{q}{t}$$ I = Current, expressed in amperes, A   q = Charge, expressed in coulombs, C   t = Time, expressed in seconds, sec

Example:

A current of 4.0 A passes for 30 seconds through an electrolytic cell that contains a aqueous solution of silver nitrate. What mass of silver will be deposited at the cathode in that time?

Solution:

$$0.13 \text{ g Ag} = 4.0 \text{ A} \times 30 \text{ sec} \times \frac{C}{A \text{ sec}} \times \frac{1.0 \text{ mol e}^-}{96{,}485 \text{ C}} \times \frac{107.868 \text{ g AG}}{1.0 \text{ mol e}^-}$$

CHAPTER 5

Equations and Stoichiometry

EQUATIONS AND STOICHIOMETRY

BALANCING CHEMICAL EQUATIONS

- Balanced chemical reactions are an artifact of the Law of Conservation of Mass. Matter is neither created nor destroyed during a chemical change.

- The same number of like atoms must exist both as reactants and products.

- To balance reactions, alter only the coefficients; do not change the formula.

- In general, when balancing reactions, balance the hydrogen and oxygen atoms last.

- For a change in oxidation number that is not a combustion reaction, see the next section, "Balancing Oxidation-Reduction reactions."

- When faced with an unbalanced reaction that involves the combustion of a hydrocarbon, look for the following:

1. First match the number of carbons in the hydrocarbon in the reactants by adjusting the coefficient of carbon dioxide in the product.

   Example:

   $\underline{C_3}H_8 + O_2 \rightarrow \underline{3C}O_2 + H_2O$

   Three carbon atoms on both sides.

2. Then balance the hydrogen atoms by adding a coefficient to water in the product so that the hydrogen atoms equal the number of hydrogen atoms in the reactant hydrocarbon.

   Example:

   $C_3\underline{H_8} + 5O_2 \rightarrow 3CO_2 + \underline{4H_2}O$

   Eight hydrogen atoms on both sides.

3. Finally, add up the oxygen atoms in the products (carbon dioxide and water) and adjust the coefficient for the oxygen molecule in the reactants so that all oxygen atoms balance.

Example:

$$C_3H_8 + \underline{5O_2} \rightarrow \underline{3}CO_{\underline{2}} + \underline{4}H_2\underline{O}$$

Ten oxygen atoms on both sides.

BALANCING OXIDATION-REDUCTION REACTIONS

- Balancing oxidation-reduction reactions requires balancing both mass and charge. Not only must the same number of each kind of atom exist on both sides of the yield sign, but the number of electrons lost in oxidation must equal the number of electrons gained during reduction.

- Balancing oxidation-reduction reactions is NOT the result of trial-and-error. The following is a well-established method that allows you to get the right answer the first time, every time.

1. Establish oxidation numbers for each atom.

2. Create two "half-reactions"; one that shows the atoms oxidized, the other that shows the atoms reduced.

3. Show how many electrons are given up in the oxidation half-reaction, and how many electrons are used in the reduction half-reaction.

4. Multiply the coefficients of each species in one or both half-reactions so that the number of electrons lost in oxidation equals the number of electrons gained in reduction.

5. Balance the mass of elements that do not change in oxidation state.

6. Recombine half-reactions.

BIG hints:

1. Try to balance the major common ions that do not have an atom that changes in oxidation state, such as the nitrate ion, as an entire unit.

2. If the reaction occurs in an acidic solution, balance the oxygen before the hydrogen. Then add the number of water molecules on the other side to balance the oxygen. Balance the remaining hydrogen atoms by adding H^+ to one side of the equation.

3. In basic solution, balance as though in an acidic solution, and then "neutralize" each H^+ by adding an equal number of OH^- ions to both sides.

Example:

Balance the following reaction in terms of both mass and charge. The reaction occurs in an acidic solution.

$$Fe(NO_3)_2 + HNO_3 + KMnO_4 \rightarrow Fe(NO_3)_3 + Mn(NO_3)_2 + H_2O + KNO_3$$

Solution:

1. **Establish oxidation numbers for each atom.**

   For now, don't worry about atoms or polyatomic ions that do not change in oxidation state.

   $$Fe(NO_3)_2 + HNO_3 + KMnO_4 \rightarrow Fe(NO_3)_3 + Mn(NO_3)_2 +$$

   +2   +7   +3   +2

   $$H_2O + KNO_3$$

2. **Create two "half-reactions"; one that shows the atoms oxidized, the other that shows the atoms reduced.**

3. **Show how many electrons are given up in the oxidation half-reaction, and how many electrons are used in the reduction half-reaction.**

   Oxidation: $Fe^{+2} \rightarrow Fe^{+3} + 1e^-$

   Reduction: $5e^- + Mn^{7+} \rightarrow Mn^{2+}$

4. **Multiply the coefficients of each species in one or both half-reactions so that the number of electrons lost in oxidation equals the number of electrons gained in reduction.**

   Oxidation: $\underline{5}Fe^{+2} \rightarrow \underline{5}Fe^{+3} + \underline{5}e^-$

   Reduction: $\underline{5}e^- + Mn^{7+} \rightarrow Mn^{2+}$

5. **Recombine half-reactions.**

6. **Balance the mass of elements that do not change in oxidation state.**

Hints:

* may want to balance polyatomic atoms as a single unit.

* Be sure to not change the coefficients of the atoms that change in oxidation state.

* If the reaction occurs in an acidic solution, balance the oxygen before the hydrogen. Then add the number of water molecules on the other side to balance the oxygen. Balance the remaining hydrogen atoms by adding H^+ to one side of the equation.

$$5\ Fe(NO_3)_2 + 8\ HNO_3 + KMnO_4 \rightarrow 5\ Fe(NO_3)_3 + Mn(NO_3)_2 + 4\ H_2O + KNO_3$$

MOLE CONVERSIONS

- The factor-label method is useful in solving mole conversions.

- Conversion factors are ratios in which the numerator is equal, by definition, to the denominator. For example, if there are 12 inches to 1 foot, then that definition can be written as either

$$\frac{12\ \text{inches}}{1\ \text{foot}} \quad \text{or} \quad \frac{1\ \text{foot}}{12\ \text{inches}}$$

in such a manner that will cancel out units of given information and result in units of the answer.

- The numbers used should all be significant figures, and the answer should contain no more significant figures than the least number of significant figures of the conversions and given information.

PARTICLES

- There are 6.02×10^{23} (Avogadro's number) particles in a mole. This value can be used to determine the number of particles in a mole of atoms or molecules.

Example:

How many atoms are in 0.2 moles of atoms?

Solution:

$$0.2\ \cancel{\text{mol atoms}} \times \frac{6.02 \times 10^{23}\ \text{atoms}}{1.0\ \cancel{\text{mol atoms}}} = 1.2 \times 10^{23}\ \text{atoms}$$

MASS

- Atomic and molecular weights can be used to convert moles to grams, and vice versa.

  Example:

  What is the mass of 4.20 moles of carbon atoms?

  Solution:

  $$4.20 \; \cancel{\text{mol atoms}} \times \frac{12.011 \text{ grams carbon}}{1.0 \; \cancel{\text{mol atoms}}} = 50.45 \text{ grams carbon}$$

GASES

- While stoichiometry problems can result in answers in moles, the ideal gas law can be used to convert that number of moles to some unknown variable in the ideal gas law equation.

- At STP (standard temperature and pressure), 1.0 mole of gas occupies 22.4 liters. This can be used as a conversion factor in solving stoichiometric equations.

  Example:

  What is the volume of 8.26 moles of gas at STP?

  Solution:

  $$8.26 \; \cancel{\text{moles gas}} \times \frac{22.4 \text{ liters gas}}{1.0 \; \cancel{\text{moles gas}}} = 185 \text{ liters gas}$$

SOLUTIONS

- Stoichiometric problems that involve solutions usually require a molar solution conversion at some point to get the answer, where

  $$\textbf{Molarity} = \textbf{M} = \frac{\textbf{moles solute}}{\textbf{Liters solution}}$$

  Example:

  How many moles of glucose exist in 1.4 liters of a 0.20-molar solution of glucose?

Solution:

$$1.4 \; \cancel{\text{liter solution}} \; \times \; \frac{0.20 \text{ moles glucose}}{1.0 \; \cancel{\text{liter solution}}} = 0.28 \text{ moles glucose}$$

STOICHIOMETRY PROBLEMS

- Stoichiometry problems refer to a type of question in which the student is given an amount of reactant or product in a chemical reaction, and then asked to find the corresponding amount of reactant or product.

- Stoichiometry problems can all be solved using the following steps:

  1. Balance the reaction.
  2. Write down the units of the answer next to an equals sign "=".
  3. On the other side of an equals sign, write down the number and units of the given information.
  4. Use a ratio to convert the given information into moles.
  5. For the next ratio, use the balanced reaction to relate the number of moles of the given substance to the number of moles of the substance in the answer.
  6. Convert the number of moles of the substance in the answer to the actual units of the answer. Be sure that your answer has the correct units, and the same units as those that remain on the other side of the equals sign after the other units are canceled.

  Example:

  In the reaction,

  $$N_2 \,(g) + H_2 \,(g) \rightarrow NH_3 \,(g)$$

  What mass of ammonia will be produced when 10.0 grams of nitrogen gas reacts completely?

  Solution:

  1. **Balance the reaction**

     $$N_2 \,(g) + 3 \, H_2 \,(g) \rightarrow 2 \, NH_3 \,(g)$$

2. **Write down the units of the answer next to an equals sign "=".**

   x grams NH_3 =

3. **On the other side of an equals sign, write down the number and units of the given information.**

   x grams NH_3 = 10.0 g N_2

4. **Use a ratio to convert the given information into moles.**

   x grams NH_3 = 10.0 $\cancel{\text{g } N_2}$ × $\dfrac{1 \text{ mol } N_2}{28.0 \cancel{\text{ g } N_2}}$

5. **For the next ratio, use the balanced reaction to relate the number of moles of the given substance to the number of moles of the substance in the answer.**

   x grams NH_3 = 10.0 $\cancel{\text{g } N_2}$ × $\dfrac{1 \text{ mol } N_2}{28.0 \cancel{\text{ g } N_2}}$ × $\dfrac{2 \text{ mol } NH_3}{1 \text{ mol } N_2}$

6. **Convert the number of moles of the substance in the answer to the actual units of the answer. Be sure that your answer has the correct units, and the same units as those that remain on the other side of the equals sign after the other units are canceled.**

   12.15 grams NH_3 = 10.0 $\cancel{\text{g } N_2}$ × $\dfrac{1 \cancel{\text{ mol } N_2}}{28.0 \cancel{\text{ g } N_2}}$ × $\dfrac{2 \cancel{\text{ mol } NH_3}}{1 \cancel{\text{ mol } N_2}}$

   × $\dfrac{17.01 \text{ g } NH_3}{1 \cancel{\text{ mol } NH_3}}$

Example that involves numbers of particles:

What number of hydrogen molecules react with 2.0 moles of nitrogen gas in the reaction,

N_2 (g) + 3 H_2 (g) ↔ 2 NH_3 (g)

Solution:

2.0 mol N_2 × $\dfrac{3 \text{ mol } H_2}{1 \text{ mol } N_2}$ × $\dfrac{6.02 \times 10^{23} \text{ molecules}}{1 \text{ mol } H_2 \text{ molecules}}$ =

3.6 × 10^{24} molecules

Example that involves mass:

What mass of hydrogen gas is used to completely react with 2.0 mol of nitrogen gas in the following reaction,

$$N_2 \, (g) + 3 \, H_2 \, (g) \leftrightarrow 2 \, NH_3 \, (g) \, ?$$

Solution:

$$2.0 \text{ mol N}_2 \times \frac{3 \text{ mol H}_2}{1 \text{ mol N}_2} \times \frac{2.016 \text{ grams H}_2}{1 \text{ mol H}_2} = 12.1 \text{ grams H}_2$$

Example that involves a gas:

What volume of ammonia gas is produced at STP when 12.6 g of hydrogen gas reacts completely in the following reaction,

$$N_2 \, (g) + 3 \, H_2 \, (g) \leftrightarrow 2 \, NH_3 \, (g) \, ?$$

Solution:

$$12.6 \text{ g H}_2 \times \frac{1 \text{ mol H}_2}{2.016 \text{ g H}_2} \times \frac{2 \text{ mol NH}_3}{3 \text{ mol H}_2} \times \frac{22.4 \text{ liters NH}_3}{1 \text{ mol NH}_3}$$

$$= 93.3 \text{ liters NH}_3$$

Example that involves a solution:

What volume of 0.10 M HCl is needed to completely neutralize 0.20 moles of sodium acetate in the following reaction,

$$H_2O + C_2H_3O_2^- \, (aq) \leftrightarrow HC_2H_3O_2 \, (aq) + OH^- \, (aq) \, ?$$

Solution:

$$0.20 \text{ mol C}_2\text{H}_3\text{O}_2^- \times \frac{1 \text{ mol HCl}}{1 \text{ mole C}_2\text{H}_3\text{O}_2^-} \times \frac{1.0 \text{ liter solution}}{0.10 \text{ mol HCl}}$$

$$= 2.0 \text{ liter solution}$$

LIMITING REACTANTS

- Limiting reactant problems are stoichiometric questions that ask you to calculate the amount of product created when given a specific amount of more than one reactant.

- Limiting reactant problems are solved like any other stoichiometric problem. However, you must go through the stoichiometric calculation with both sets of given information.

- The limiting reactant is the reactant that runs out first as the chemical reaction proceeds. Therefore, it is the reactant that produces the least amount of product.

Example:

In the following reaction, a 12.8 L sample of CO at 2.0 atm and 300 K is combined with 6.33 g of Fe_2O_3. How many liters of carbon dioxide are formed at the same temperature, and which is the limiting reactant?

Fe_2O_3 (s) + 3 CO (g) → 2 Fe (s) + 3 CO_2 (g)

Solution:

(i) $\quad 12.8 \text{ L CO} \times \dfrac{3 \text{ L CO}_2}{3 \text{ L CO}} = 12.8 \text{ L CO}_2$

(ii) $\quad 6.33 \text{ g Fe}_2O_3 \times \dfrac{1 \text{ mol Fe}_2O_3}{159.7 \text{ g Fe}_2O_3} \times \dfrac{3 \text{ mol CO}_2}{1 \text{ mol Fe}_2O_3} \times$

$\dfrac{(.082 \text{ L atm})(300 \text{ K})}{2.0 \text{ atm K mol}} = 1.46 \text{ L CO}_2$

Therefore, iron oxide is the limiting reactant and 1.46 L of carbon dioxide is formed.

EMPIRICAL FORMULAS AND PERCENT COMPOSITION

- The empirical formula is the simplest ratio of atoms in a molecule. The empirical formula may or may not be the same as the molecular formula. The molar mass is the information that is needed to convert from the empirical formula to the molecular formula.

- Mass-mole conversions are necessary in finding empirical formulas from experimental data using the following steps:

1. Change the "%" to grams by assuming that you have 100 grams of material.

2. Convert from mass into grams for all atoms involved.

3. Divide the moles of each atom by the value that represents the smallest number of moles of atoms.

Example:

(a) What is the empirical formula of the hydrocarbon that contains 85.7% carbon?

(b) Vapor pressure calculations determine the molar mass of the compound to be 28.0 g/mol. What is the molecular formula of the compound?

Solution:

(a) $85.7 \text{ g C} \times \dfrac{1.0 \text{ mol C}}{12.011 \text{ g C}} = 7.135 \text{ mol C}$

$14.3 \text{ g H} \times \dfrac{1.0 \text{ mol H}}{1.008 \text{ g H}} = 14.186 \text{ mol H}$

Empirical formula: $C_{7.135}H_{14.186}$, or CH_2

(b) Since the molar mass is twice the mass of the empirical formula, then the molecular formula is twice the empirical formula; C_2H_4.

CHAPTER 6

Equilibrium

CHAPTER 6

EQUILIBRIUM

LE CHATELIER'S PRINCIPLE

- **Le Chatelier's principle** states that when a system at equilibrium is disturbed by a change in pressure, temperature, or the amount (concentration) of product or reactant, the reaction will shift to minimize the change and establish a new equilibrium.

- **Change in concentration:** Adding products to a reaction the equilibrium will shift the reaction to produce reactants; adding reactants to a reaction at equilibrium will shift the reaction to produce products. See the section on the reaction quotient, Q, below.

- **Change in temperature:** An increase in temperature causes the equilibrium to shift to use up the added heat. For example, when heat is added to an exothermic reaction, it will shift to the left to use up the heat. An endothermic reaction will shift to the right to use up heat when heat is added.

- **Change in pressure:** An increase in pressure causes the equilibrium to shift in the direction that produces the fewest number of gas moles. For example, in the reaction that dissolves a gas into a liquid, increasing the pressure on the system will cause the equilibrium to shift to produce more dissolved gas.

- **Addition of a catalyst or an inert gas** will not cause the equilibrium to shift; the amounts of reactants and products would remain unchanged.

EQUILIBRIUM CONSTANTS

- Equilibrium constants are ratios.

  For the reaction, $\mathbf{aA + bB \leftrightarrow cC + dD,}$

The ratio of the product concentrations, raised to their stoichiometric coefficients, to the reactant concentrations, raised to their stoichiometric coefficients, is the equilibrium constant, K_{eq}. This is also called the **law of mass action**.

$$K_{eq} = \frac{[C]^c\,[D]^d}{[A]^a\,[B]^b}$$

- Pure substances, such as water or solids, do not show up in the equilibrium expression; only molar solutions or, as with K_p, gaseous pressures.

- The **equilibrium constant for a multi-step process** is equal to the product of the equilibrium constants for each step.

  Example: For a set of three reactions that add to equal a total reaction,

  $$K_{total} = K_1 \times K_2 \times K_3$$

- The **equilibrium constant for a reverse reaction** is the inverse of the equilibrium constant for a forward reaction.

  Example:

  $$K_{reverse} = \frac{1}{K_{forward}}$$

- There are different equilibrium constants for different types of reactions.

Type of Reaction	Equilibrium Constant
Reaction in solution; reactants and products expressed as a concentration in moles/liter.	K_c
Gaseous reaction, reactants and products expressed in units of pressure. $N_2\ (g) + 3\ H_2\ (g) \leftrightarrow 2\ NH_3\ (g)$	K_p
The dissociation of water $H_2O \leftrightarrow H^+\ (aq) + OH^-\ (aq)$	K_w
Reactions that produce a proton (H^+) from a Brønsted-Lowry acid. e.g.: $HC_2H_3O_2\ (aq) \leftrightarrow H^+\ (aq) + C_2H_3O_2^-\ (aq)$	K_a
Reactions that produce a hydroxide ion (OH^-) from a Brønsted-Lowry base. e.g.: $H_2O + C_2H_3O_2^-\ (aq) \leftrightarrow HC_2H_3O_2\ (aq) + OH^-\ (aq)$	K_b
Reactions that produce dissolved ions in aqueous solution from a solid. e.g.: $PbI_2\ (s) \leftrightarrow Pb^{2+}\ (aq) + 2\ I^-\ (aq)$	K_{sp}

REACTION QUOTIENT, Q

- The reaction quotient, Q, can be used to calculate the direction and degree to which a reaction will shift when new products or reactants are added (Le Chatelier's principle).

- The reaction quotient for a reaction is found using the same ratio as the equilibrium constant, but at non-equilibrium conditions.

  For the reaction, $\mathbf{aA + bB \leftrightarrow cC + dD}$

  $$Q = \frac{[\mathbf{C}]^c\,[\mathbf{D}]^d}{[\mathbf{A}]^a\,[\mathbf{B}]^b}$$

- If the reaction quotient is greater than the equilibrium expression, there are more products than there would be at equilibrium. By Le Chatelier's principle, the reaction will shift toward equilibrium by using products and producing more reactants.

  When Q > K, reaction proceeds to the left toward reactants.

  When Q < K, reaction proceeds to the right toward products.

  When Q = K, reaction is at equilibrium.

  Example:

  In the following reaction,

  $N_2\ (g) + 3\ H_2\ (g) \leftrightarrow 2\ NH_3\ (g)$

  The K_c is 5.9×10^{-2}. The molar concentrations of each reactant and product are: $[N_2] = 0.40$ M, $[H_2] = 0.80$ M, and $[NH_3] = 0.20$ M. Which direction will the reaction go as it begins to establish equilibrium?

  Solution:

  $$Q = \frac{(.20)^2}{(.4)\,(.8)^3} = 0.195$$

  Q > K, therefore the reaction goes to the left (toward the reactants).

EQUILIBRIUM CONSTANTS FOR GASEOUS REACTIONS

- Concentrations of reactants and products in the gas phase may be expressed in units of molarity (moles/liter) or partial pressures (atm, Pa, mmHg).

- The ideal gas law (P = NRT) is used to convert between the equilibrium constant expressed in gas moles/liter (K_c) to the equilibrium constant expressed in gas pressures (K_p); see below.

- The equilibrium expression for K_p only contains those species that are in the gas phase. For example, the K_p for the vaporization of water from liquid is simply the equilibrium vapor pressure of water.

GAS EQUILIBRIUM CONSTANT

$$K_p = K_c \, (RT)^{\Delta n}$$

K_p = Equilibrium constant, partial pressures

K_c = Equilibrium constant, molar concentrations

R = ideal gas constant, 0.082 L atm / K mol

T = Absolute temperature, K

Δn = (Moles of gas in products) − (Moles of gas in reactants)

Example:

The value of K_p is 8.3×10^{-3} at 700 K for the reaction,

$N_2 \, (g) + 3 \, H_2S \, (g) \leftrightarrow 2 \, NH_4HS \, (g)$

What is the value of K_c?

Solution:

$$\frac{K_p}{(RT)^{\Delta n}} = K_c, \text{ where } \Delta n = 1$$

$$\frac{K_p}{(R)(T)} = K_c = (8.3 \times 10^{-3})/(0.082)(700) = 1.46 \times 10^{-4}$$

EQUILIBRIUM CONSTANTS FOR REACTIONS IN SOLUTION

The Dissociation of Weak Acids and Bases

- The water dissociation constant, $K_w = 10^{-14} = K_a K_b = [H^+] [OH^-]$. This relationship can be used to freely convert between all these concentrations and equilibrium constants.

- Equilibrium constants, K_a and K_b, can be used to calculate the pH of a weak acid or base solution.

  Example:

  What is the pH of a 0.10 molar solution of acetic acid, $HC_2H_3O_2$, which has a $K_a = 1.8 \times 10^{-5}$?

  Solution:

  1. Write the equation of the reaction that occurs when the weak acid or base is put in water, and its corresponding equilibrium expression.

  $$HC_2H_3O_2 \ (aq) \leftrightarrow H^+ \ (aq) + C_2H_3O_2^- \ (aq)$$

  $$K_a = \frac{[H]^+ \ [C_2H_3O_2^-]}{[HC_2H_3O_2]}$$

  2. When the weak acid is placed in water, some number of moles/liter (x) will dissociate. For each mole of weak acid that dissociates, one mole of the proton and one mole of the weak base will be formed.

  $$HC_2H_3O_2 \ (aq) \leftrightarrow H^+ \ (aq) + C_2H_3O_2^- \ (aq)$$

	$HC_2H_3O_2$	H^+	$C_2H_3O_2^-$
Before equilibrium:	0.10M	0	0
After equilibrium:	0.10 − x	x	x

  3. Write the equilibrium expression in terms of the number of dissociated moles, x, which also happens to be the molar concentration of protons.

  $$K_a = \frac{(x^2)}{(0.10 - x)} = 1.8 \times 10^{-5}$$

  4. Assume that the amount of x that dissociates is small relative to the amount of weak acid, so neglect the x portion of the denominator while solving the x in the numerator. (Hint: this works when there

is a small level of dissociation. To not make this assumption would require the quadratic equation to solve the problem.)

$$x^2 = 1.8 \times 10^{-5} \times 0.10$$

$$x = \sqrt{1.8 \times 10^{-6}} = [H^+]$$

$$[H^+] = 1.34 \times 10^{-3}$$

$$pH = -\log [H^+]$$

$$pH = -\log [1.34 \times 10^{-3}]$$

$$pH = 2.87$$

Solubility Product Constants

- The solubility product constant (K_{sp}) refers to the product of the molar concentration of soluble ions that exists at the saturation point of the solution.

- K_{sp} refers to ionic compounds that are considered soluble. However, to some degree, some miniscule amount of ions dissolve from "insoluble" compounds. The K_{sp} can tell you the degree to which those ions have dissolved.

- There are two types of problems you will see that involve K_{sp}:

  1. Find the molar solubility of a compound from K_{sp}, or vice versa.

  2. Determine whether or not a precipitate will form when two solutions are added together.

- The molar solubility of a compound can be calculated from K_{sp} in the following example.

  Example:

  The K_{sp} for barium carbonate ($BaCO_3$) is 2.58×10^{-9}. How many moles of $BaCO_3$ can dissolve in 1.0 liter of solution?

  Solution:

  $$x^2 = 2.58 \times 10^{-9}$$

  $$x = 5.08 \times 10^{-5}$$

- You can find the K_{sp} from the molar solubility in the following example.

Example:

The molar solubility for copper sulfide (CuS) is 9.2×10^{-23} M. What is the solubility product constant for CuS?

Solution:

$K_{sp} = (x)^2 = (9.2 \times 10^{-23})^2$

$K_{sp} = 8.5 \times 10^{-45}$

COMMON ION EFFECT

- The common ion effect occurs when the ion of one equilibrium exists also in another equilibrium. One concentration satisfies both equilibria.

- The following example shows how to calculate the molar solubility of a compound in a solution where a common ion is already dissolved in the solution.

Example:

The K_{sp} for magnesium hydroxide ($Mg(OH)_2$) is 8.9×10^{-12}. What is the molar solubility of $Mg(OH)_2$ in a solution that is pH 4 ?

Solution:

$K_{sp} = [Mg^{2+}] [OH^-]^2$ (where $[OH^-] = 10^{-10}$ M)

molar solubility $= [Mg^{2+}] = K_{sp} / 10^{-20} = 8.9 \times 10^8$ M (very soluble)

BUFFERS

- A buffer is an aqueous combination of a weak acid and its conjugate base, or a weak base and its conjugate acid.

- A buffer solution resists a change in pH when new H^+ or OH^- ions are added. The H^+ ions react with the weak base to form water; the OH^- ions react with the weak acid to form water.

- Maximum buffer capacity occurs when pH = pK_a for the weak acid. However, the buffering capacity will be exhausted if either the weak acid or the weak base is used up.

- Buffers are best described by the Henderson-Hasselbach equation, which is derived by solving for $[H^+]$ in the equilibrium expression, then taking the negative log of each side of the equation.

THE HENDERSON-HASSELBACH EQUATION

$$pH = pK_a + \log\frac{[A^-]}{[HA]}$$

$[A^-]$ = molar concentration of conjugate base

$[HA]$ = molar concentration of weak acid

pH = pH of the solution

pK_a = $-\log K_a$ of the weak acid

Example:

What is the pH of the buffer solution created by combining 100 mL of 0.2 M acetic acid and 400 mL of 0.10 M sodium acetate? ($K_a = 1.8 \times 10^{-5}$)

Solution:

$$[H^+] = \frac{K_a[HC_2H_3O_2]}{[C_2H_3O_2^-]} = \frac{(1.8 \times 10^{-5})\,(.1\text{ L})\,(.2\text{ M}/.5\text{ L})}{(.4\text{ L})\,(.1\text{ M}/.5\text{ L})} =$$

9.0×10^{-6} M $= [H^+]$

pH $= -\log[H^+]$

pH $= -\log[9.0 \times 10^{-6}]$

pH $= 5.0$

CHAPTER 7

Kinetics

CHAPTER 7

KINETICS

REACTION RATE

- Kinetics determines the speed of a reaction, and is dependent on the mechanism by which reactants turn into products.

- Reaction rate is based on the rate of appearance of a product or disappearance of a reactant, and is expressed as a change in concentration over time.

- Reaction rates are determined experimentally by measuring concentrations.

- Reaction rates increase by increasing concentration or by increasing the temperature.

- Reaction rates may also increase by increasing the surface area of the reactant—which is the same as increasing the concentration of possible participants in the reaction, or by adding a catalyst.

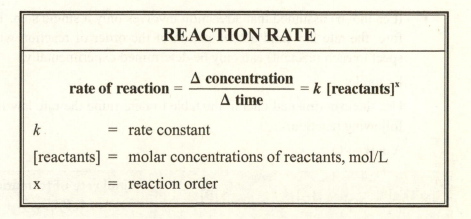

REACTION RATE

$$\text{rate of reaction} = \frac{\Delta \text{ concentration}}{\Delta \text{ time}} = k \ [\text{reactants}]^x$$

k = rate constant

[reactants] = molar concentrations of reactants, mol/L

x = reaction order

RATE LAW AND REACTION ORDER

- The **rate law** describes the rate of the reaction as a function of a rate constant, which is dependent on the temperature and the concentrations of the reactants.

- All rate laws take the form of **rate = k [reactants]^x**, where k is the rate constant, [reactants] refers to the molar concentration of reactants, and the exponent, x, is the reaction **order**.

- For all **single-step reactions**, the rate law for a particular step can be surmised from the balanced equation for that step because the rate of the forward reaction is proportional to the concentration of available reactants.

  Example:

  Total reaction (single step): $A \rightarrow B + C$ rate = k [A]

    (first order)

  Or

  Total reaction (single step): $A + A \rightarrow B$ rate = k [A]^2

    (second order)

  Or

  Total reaction (single step): $A + B \rightarrow C$ rate = k [A] [B]

    (second order overall)

- It cannot be assumed that a reaction involves only a single step. Therefore, the rate law (and determination of the order of reaction with respect to each reactant) can only be determined experimentally.

  Example:

  Use the experimental data in the table to determine the rate law for the following reaction,

  $A + B \rightarrow C$

Trial	[A]	[B]	Initial rate of formation of C (M/sec)
1	0.10	0.40	0.02
2	0.10	0.80	0.04
3	0.20	0.80	0.16

Solution:

When doubling [B] between the first and second trial (holding [A] constant), the reaction rate also doubles. Therefore, the rate of the reaction is *first order* with respect to [B]. When doubling [A] between the second and third trial (holding [B] constant), the reaction rate quadruples. Therefore, the rate changes as a square of a change in [A]; the reaction rate is *second order* with respect to [A]. The overall rate law of the reaction is the combination of the two orders: rate = k [A]^2 [B].

• The integrated rate laws for the different orders of reaction can be used to calculate the concentrations of reactants at specific times. The following table summarizes the integrated rate laws, straight line plots, and methods of calculating half-life for the different reaction orders.

	Zero Order Reactions	First Order Reactions	Second Order Reactions
Rate law	k	k [A]	k [A]^2
Integrated rate law	[A] = $-kt$ + [A$_{initial}$]	ln[A] = $-kt$ + ln[A$_{initial}$]	$\frac{1}{A} = kt + \frac{1}{[A_{initial}]}$
Plot for straight line	[A] vs. t	ln [A] vs. t	$\frac{1}{[A]}$ vs. t
k and slope of line	Slope = $-k$	Slope = $-k$	Slope = k
Half-life	[A$_{initial}$] / 2k	0.693 / k	1 / k [A$_{initial}$]

KINETICS AND EQUILIBRIUM

• Equilibrium occurs when the rate of the forward reaction equals the rate of the reverse reaction.

• The relationship between the equilibrium constant, K, and the rate constants, k_f and k_r, can be found by setting the rate law of the forward reaction equal to the rate law of the reverse reaction, then isolating the concentrations on one side of the equation and the rate constants on the other.

KINETICS AND EQUILIBRIUM

At equilibrium, $k_{forward}$ [**reactants**]^y = $k_{reverse}$ [**products**]^x

$$\mathbf{K}_{eq} = \frac{k_{forward}}{k_{reverse}} = \frac{[\mathbf{products}]^x}{[\mathbf{reactants}]^y}$$

TEMPERATURE AND REACTION RATES

- Reaction rate depends on temperature since molecules at greater temperature have greater kinetic energy and collide both more frequently and with greater force.

- When collisions occur with sufficient energy, or the **activation energy**, then a reaction will occur.

- The rate constant (which is proportional to reaction rate) is different at different temperatures. The rate constant will double (approximately) for every 10 K increase in temperature.

TEMPERATURE, RATE CONSTANT, AND ACTIVATION ENERGY

$$\ln k = \frac{-\mathbf{Ea}}{\mathbf{RT}} + \ln [\mathbf{A}]$$

Where:

k = rate constant (proportional to reaction rate)

Ea = Activation energy, J

[A] = molar concentration of reactant, A, mol/L

R = Ideal gas constant, 8.31 J/K mol

T = temperature, K

ACTIVATION ENERGY AND CATALYSTS

- The **activation energy** is the amount of energy needed to cause a reaction to occur.

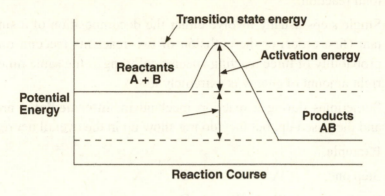

$$\Delta E = \Sigma\, E \text{ products} - \Sigma\, E \text{ reactants}$$

- **Catalysts** increase the rate of a reaction by facilitating a faster pathway, or mechanism, without being used up itself during the reaction. Therefore, catalysts do not appear in a balanced chemical equation.

- The new mechanism of reaction created by a catalyst has a lower activation energy.

- In living systems, catalysts are frequently used to speed up reactions that—without the catalyst—might be too slow to be useful for the organism.

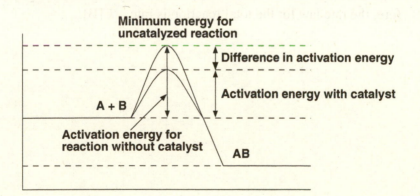

REACTION MECHANISMS

- Most chemical reactions involve more than one step. The combination of steps is called the **mechanism** of reaction, and must add to yield the total reaction.

- Single steps usually involve either the decomposition of a single reactant (first order) or the collision of two reactants (second order). The probability of three reacting species colliding at the same time with the right amount of energy is extremely low.

- Sometimes during a reaction mechanism, **intermediates** are formed and then used up; but they do not show up in the overall net reaction.

  Example:

Step one:	A		$\rightarrow$	C + X	
Step two:	B + X	$\rightarrow$	D		
Total reaction:	A + B	$\rightarrow$	C + D		

- The slowest step in the set of reactions in a mechanism determines the overall rate of the reaction, so it is called the **rate-determining step**.

  Example:

Step one:	A	$\rightarrow$	C	
Step two:	B	$\rightarrow$	D	slow step (rate-
Total reaction:	A + B	$\rightarrow$	C + D	determining step)

  If the total reaction were a single step, the total rate law would be

  Rate = k [A] [B]. However, in this multistep mechanism, the total rate depends only on the second step, which only depends on [B]. Therefore, the rate law for the total reaction is rate = k [B].

CHAPTER 8

Thermodynamics

CHAPTER 8

THERMODYNAMICS

STATE FUNCTIONS

What Are State Functions?

- The state of a system is defined by conditions of pressure, temperature, and number of moles of a substance.

- State functions depend only on initial and final states of a system. Thermodynamics deals with initial and final states.

- State functions are not dependent on how one gets from the initial state to the final state. The mechanism of getting from one state to another is represented by the kinetics of a reaction.

Types of State Functions

- Entropy (S), enthalpy (H), and free energy (G) are state functions.

- **Entropy** is a measure of disorder, measured in J/K.

- **Enthalpy** is energy released or absorbed by the reaction, measured in J.

- **Free energy** is the energy available to do work, measured in J.

- Most thermodynamics at this level deals with changes in entropy, enthalpy, and free energy between an initial state and a final state. (ΔS, ΔH, and ΔG, respectively.)

Standard State Conditions

- Standard state conditions for state functions are indicated by a "°".

  Example: $\Delta S°$ represents the standard change in entropy; $\Delta H°$ represents the standard change in enthalpy, and $\Delta G°$ represents the standard change in free energy.

- Standard conditions are a very specific set of conditions, which include

    1. Gases are at 1 atmosphere pressure.

    2. Liquids and solids are pure.

    3. Solutions are at 1molar concentration.

    4. The temperature is 25° C, or 298 K.

- Do not confuse thermodynamic standard conditions, where the temperature is 298 K, with standard temperature and pressure (STP) for gases, where the temperature is 273 K.

FIRST LAW AND ENTHALPY

First Law of Thermodynamics

- **The first law of thermodynamics** is commonly referred to as the law of Conservation of Energy.

- More specifically, the first law states that the change in internal energy is equal to the difference between the energy supplied to the system as heat and the energy removed from the system as work performed on the surroundings.

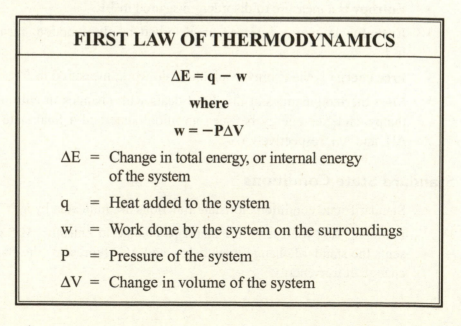

FIRST LAW OF THERMODYNAMICS

$$\Delta E = q - w$$

where

$$w = -P\Delta V$$

ΔE = Change in total energy, or internal energy of the system

q = Heat added to the system

w = Work done by the system on the surroundings

P = Pressure of the system

ΔV = Change in volume of the system

Example:

An ideal gas absorbs 100 L atm of heat energy, which expands against a pressure of 8.0 atm. The total energy of the system is -20.0 L atm. How much does the volume change during the expansion?

Solution:

$$\Delta E = q - w = q + (P\Delta V)$$

$$\Delta V = \frac{E - q}{P} = \frac{-20.0 \text{ L atm} - 100 \text{ L atm}}{8.0 \text{ atm}} = 15 \text{ L}$$

Enthalpy

- **Enthalpy** is a state function.

- The **enthalpy change** in a reaction (ΔH) is the difference between the enthalpy contained in the products and the enthalpy contained in the reactants.

- **Exothermic reactions** give off heat energy, $\Delta H_{reaction} < 0$. The products of the reaction have less enthalpy than the reactants.

- **Endothermic reactions** take in heat energy, $\Delta H_{reaction} > 0$. The products of the reaction have more enthalpy than the reactants.

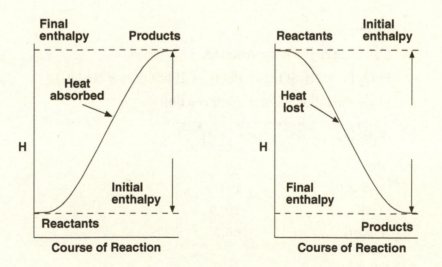

ENTHALPY CHANGE
DURING A REACTION

$$\Delta H_{reaction} = \Sigma H_{products} - \Sigma H_{reactants}$$

Heats of Formation, Heats of Reaction

- The **standard heat of formation** ($\Delta H°_f$) is the change of enthalpy for the reaction that forms a compound from its pure elements under standard conditions.

- The standard heat of formation ($\Delta H°_f$) for a pure element at standard conditions is zero.

- Standard heats of formation can be used to estimate the $\Delta H°$ of any reaction

---

$\Delta H°$ AND HEATS OF FORMATION

$$\Delta H° = \Sigma H°_f \text{ (products)} - \Sigma H°_f \text{ (reactants)}$$

Example:

Calculate $\Delta H°$ for the reaction,

$$PbO_2 \text{ (s)} + 2H_2SO_4(l) + Pb(s) \rightarrow 2PbSO_4(s) + 2H_2O(l)$$

given the following $\Delta H°f$ information.

Species	$\Delta H°_f$ (kJ)
PbO_2 (s)	−66.3
$H_2SO_4(l)$	−194.5
$PbSO_4(s)$	−219.9
$H_2O(l)$	−68.3

Solution:

$\Delta H°_{reaction} = \Sigma H°_f \text{ (products)} - \Sigma H°_f \text{ (reactants)}$

$\Delta H°_{reaction} = [(2 \times -219.9) + (2 \times -68.3)] - [(-66.3) + (2 \times -194.5)]$

$\Delta H°_{reaction} = -121.1 \text{ kJ}$

Hess's Law

- Hess's Law states the $\Delta H°$ of a reaction that is composed of multiple steps is equal to the sum of the $\Delta H°$ from each step. Hess's law is an offshoot of the first law of thermodynamics because energy must be conserved in order for the sum of the energies of component reactions to be equal to the energy of the total reaction.

Example:

	Reaction	$\Delta H_{reaction}$ (kJ/mol)
1.	$N_2 \, (g) + O_2 \, (g) \rightarrow 2\,NO$	$\Delta H_1 = 180$
2.	$2NO(g) + O_2 \, (g) \rightarrow 2\,NO_2$	$\Delta H_2 = -112$
total	$N_2 \, (g) + O_2 \, (g) \rightarrow 2\,NO_2$	$\Delta H_{total} = 68$

Bond Energies

- **Bond energies** are the amount of energy given off when bonds are formed, or the amount of energy used when bonds are broken.

- Bond energies deal with reactants and products in their gaseous state under standard conditions.

- Breaking bonds is an exothermic process; making bonds is an endothermic process.

- Heats of reaction can be estimated by finding the difference between the bond energies of the bonds made and the bond energies of the bonds broken.

- Bond energies used in this way to find heats of reaction is an example of Hess's Law.

$\Delta H°$ AND BOND ENERGIES

$$\Delta H°_{reaction} = \Sigma H°_{bond\ energies} \text{ (bonds broken)} - \Sigma H°_{bond\ energies} \text{ (bonds made)}$$

Example:

Use the following table of bond energies to approximate the change in enthalpy when one mole of propane undergoes complete combustion, according to this reaction:

$C_3H_8 + 5O_2 \rightarrow 3CO_2 + 4H_2O$

Bond	$\Delta H°_{bond}$ (kJ/mol)
C – H	413
C – C	347
O – H	467
C = O	794
O = O	495

Solution:

$\Delta H°_{reaction} = \Sigma H°_{bond}$ (bonds broken) $- \Sigma H°_{bond}$ (bonds made)

Bond broken	Bonds made
8 × C – H	8 × O – H
1 × C – C	6 × C = O
5 × O = O	

$\Delta H°_{reaction} = [(347) + (8 \times 413) + (5 \times 495)] - [(6 \times 794) + (8 \times 467)]$

$\Delta H°_{reaction} = -2374$ kJ

Calorimetry

- **Calorimetry** is the laboratory measurement of heats of reaction. The measured change in temperature of the calorimeter identifies how much heat is either absorbed or contributed to the reaction inside it.

- The heat given off by the reaction equals the heat absorbed by the calorimeter. Likewise, the heat absorbed by the reaction equals the heat given off by the calorimeter.

- **Coffee cup calorimeters** assume that all the heat of a reaction is absorbed by water in the calorimeter.

- **Bomb calorimeters** contain materials other than water that absorb the heat from the reaction. For bomb calorimeters, the heat absorbed equals the specific heat capacity of the calorimeter (in J/°C) multiplied by the change in the temperature of the calorimeter (ΔT).

CALORIMETRY

Coffee cup: $\Delta H_{reaction} = -q = -mc\Delta T$

Bomb: $\Delta H_{reaction} = -q = -C\Delta T$

q	=	heat added to (or subtracted from) calorimeter, J
m	=	mass of water in calorimeter, g
c	=	specific heat of water in calorimeter, J/ g °C
ΔT	=	temperature change of calorimeter, in °C
$\Delta H_{reaction}$	=	heat given off (or absorbed) by the reaction, J
C	=	heat capacity of the bomb calorimeter, J/°C

- Measured heats of reaction can be compared to heats of reaction that are estimated using heats of formation or bond energies.

Example:

A bomb calorimeter with a heat capacity of 10.8 J/°C is used to find the heat of combustion for a molecule. If the calorimeter increases in temperature by 82 °C during the reaction, what is the measured heat of combustion for the reaction?

Solution:

$\Delta H_{reaction} = -q = -C\Delta T$

$\Delta H_{reaction} = -10.8 \text{ J/°C} \times 82 \text{ °C} = -885.6 \text{ J}$

SECOND LAW, ENTROPY, AND FREE ENERGY

ENTROPY AND THE SECOND LAW OF THERMODYNAMICS

- **Entropy** is a measure of disorder, and carries the units of J/K.

- Entropy is a state function. The change in entropy of a reaction (ΔS) is the difference between the entropy contained in the products and the entropy contained in the reactants.

$\Delta S°$ FOR A REACTION (SYSTEM)

$$\Delta S° = \Sigma\ S°\ \textbf{(products)} - \Sigma S°\ \textbf{(reactants)}$$

- The **second law of thermodynamics** states that the entropy of the universe increases during a spontaneous reaction.

- The change in entropy of the universe is defined as the ΔS of the system (or the reaction) + the ΔS of the surroundings.

ENTROPY, ENTHALPY, AND FREE ENERGY

- The entropy of the universe is defined as the $\Delta S_{system} + \Delta S_{surroundings}$, which must increase in a spontaneous reaction, according to the second law of thermodynamics.

  — $\Delta S_{surroundings} = -\Delta H\ /\ T$, therefore

  — $\Delta S_{system} + \Delta S_{surroundings}$ can be rewritten as $\Delta S_{system} -\Delta H\ /\ T$, which must increase in a spontaneous reaction.

  — Multiplying by $-1/T$, $\Delta H - T\Delta S_{system}$ must then increase in a spontaneous reaction.

  — $\Delta H - T\Delta S_{system}$ is defined as free energy, ΔG, which is the amount of work that a reaction is able to do.

  — Therefore, for spontaneous processes, $\Delta G < 0$.

FREE ENERGY

$$\Delta G° = \Delta H° - T\Delta S°$$

$\Delta G°$ = standard free energy change of reaction, J

$\Delta H°$ = standard enthalpy change of reaction, J

T   = temperature, K

$\Delta S°$ = standard entropy change of reaction, J/K

CONDITIONS OF REACTION SPONTANEITY

$\Delta H°$	T	$\Delta S°$	$\Delta G°$	Spontaneous?
<0	Low	>0	<0	Yes
<0	High	>0	<0	Yes
>0	Low	<0	>0	No
>0	High	<0	>0	No
>0	Low	>0	>0	No
>0	High	>0	<0	Yes
<0	Low	<0	<0	Yes
<0	High	<0	>0	No

FREE ENERGY OF REACTION

Free Energy of Formation

- The free energy of formation of a substance is the $\Delta G°$ for the reaction that forms the substance from uncombined elements as they exist under standard conditions.

- The free energy is the maximum amount of heat that can be devoted to performing work by a reaction.

Example:

Use the following table of free energies to calculate the free energy for the combustion of one mole of ethanol.

$$2 C_2H_5OH \ (l) + 6O_2 \ (g) \rightarrow 4CO_2 \ (g) + 6H_2O \ (g)$$

Species	ΔG_f° (kJ/mol)
C_2H_5OH (l)	$\Delta G_f = -175$
O_2 (g)	$\Delta G_f = 0$
CO_2 (g)	$\Delta G_f = -394$
H_2O (g)	$\Delta G_f = -229$

Solution:

$\Delta G^\circ_{reaction} = \Sigma \Delta G^\circ_f \text{(products)} - \Sigma \Delta G^\circ_f \text{(reactants)}$

$\Delta G^\circ_{reaction} = [(4 \times -394) + (6 \times -229)] - (-175)$

$\Delta G^\circ_{reaction} = -2775 \text{ kJ}$

Hess's Law and Free Energies

- Hess's law works with free energies as well as enthalpies.
- The ΔG° of a reaction that is composed of multiple steps is equal to the sum of the ΔG° from each step.

	Reaction			$\Delta G_{reaction}$ (kJ/mol)
1.	N_2 (g) + 2 O_2 (g)	$\rightarrow$	2 NO_2 (g)	$\Delta G_1 = 104$
2.	2 NO_2 (g) +	$\rightarrow$	2 N_2O_4 (l)	$\Delta G_2 = 90$
total	N_2 (g) + 2 O_2 (g)	$\rightarrow$	2 N_2O_4 (l)	$\Delta G_{total} = 194$

ΔG°, K, AND E°

FREE ENERGY, EQUILIBRIUM, CELL VOLTAGE

$$\Delta G° = -nFE°_{total} = -RT \ln K$$

$\Delta G°$	=	standard free energy change of reaction at equilibrium, J
n	=	number of moles of electrons transferred in the reaction, mol
F	=	One faraday, 96,486 coulombs/mol electrons
$E°_{total}$	=	Standard cell voltage, V
R	=	Ideal gas constant, 8.31 J / mol-K
T	=	temperature, K
K	=	equilibrium constant

Example:

Given the following thermodynamic data, calculate $\Delta G°$, equilibrium constant, and $E°_{total}$ for the reaction that turns iron metal into rust ($Fe_2O_2(s)$) under standard conditions.

Species	$\Delta H°_f$ (kJ)	$S°$ (J/ K mol)
Fe_2O_3 (s)	−826	90
O_2 (g)	0	205
Fe (s)	0	28

Solutions:

(i) Find $\Delta H°$ for the reaction, $2Fe(s) + 3/2\ O_2\ (g) \rightarrow Fe_2O_3\ (s)$

$\Delta H°_{reaction} = \Sigma H°f\,(\text{products}) - \Sigma H°f\,(\text{reactants}) = -826\ kJ$

(ii) Find ΔS° for the reaction, $2Fe(s) + 3/2\ O_2\ (g) \rightarrow Fe_2O_3\ (s)$

$ΔS° = Σ\ S°\ \text{(products)} - Σ\ S°\ \text{(reactants)}$

$ΔS° = (90) - [(1.5 \times 205) + (2 \times 27)] = -271.5 = -272\ \text{J/K mol}$

(iii) Find ΔG° for the reaction, $2Fe(s) + 3/2\ O_2\ (g) \rightarrow Fe_2O_3\ (s)$

$ΔG° = ΔH° - TΔS° = -826 - (298)(-0.272) = -745\ \text{kJ}$

(iv) Find $E°_{total}$ from ΔG°

$ΔG° = -nFE°_{total}$

$E°_{total} = \dfrac{ΔG°}{nF} = 745{,}000\ \text{J} / (3)(98{,}486\ \text{C}) = 2.52\ \text{V}$

(v) Find K from ΔG°

$ΔG° = -RT\ \ln K$

$\ln K = -ΔG°/RT = 300$

$K = e^{-ΔG°/RT} = e^{300}$ (a very large number!)

CHAPTER 9

Descriptive Chemistry

CHAPTER 9

DESCRIPTIVE CHEMISTRY

PRODUCTS OF CHEMICAL REACTIONS

Precipitation Reactions

If you know the solubility rules, identifying these reactions and writing the resulting net ionic equations should be straightforward. Remember to not include spectator ions in the net ionic reaction. Only those ions that come together to form the insoluble inorganic compound should be written.

> Example: Solutions of silver nitrate and potassium iodide are mixed.
>
> Answer: $Ag^+ + I^- \rightarrow AgI$

Acid–Base Reactions

Remember that most acid–base reactions simply move a proton from one reactant to another.

1. **A strong acid neutralizes a strong base**

   The net ionic reaction for this type of acid-base neutralization is always the same. Students should memorize the strong acids and bases.

   Example: Strong hydrochloric acid is added to a solution of sodium hydroxide.

   Solution: $H^+ + OH^- \rightarrow H_2O$

2. **A strong acid neutralizes a weak base**

   The net ionic equation should depict a proton combining with the basic molecule.

   Example: Strong hydrochloric acid is added to an ammonia solution.

   Solution: $H^+ + NH_3 \rightarrow NH_4^+$

3. **A strong base reacts with a weak acid**

A proton from the weak acid combines with the hydroxide ion to form an anion and water.

Example: Solutions of sodium hydroxide and acetic acid are mixed.

Solution: $HC_2H_3O_2 + OH^- \rightarrow C_2H_3O_2^- + H_2O$

4. **A weak acid reacts with a weak base**

Keep any eye out for Lewis acid-base reactions, in which the acid accepts an electron pair from and combines with a weak base.

Example: Solutions of boron trifluoride and ammonia are mixed.

Solution: $BF_3 + NH_3 \rightarrow BF_3NH_3$

5. **A coordination-complex is formed**

This is another example of a Lewis acid-base reaction, where the metal serves as an electron-acceptor. You would recognize this if a transition metal is placed in a solution with soluble ammonia, cyanide, hydroxide, or thiocyanate ions. You may combine the metal with as many polyatomic anions as you wish; just be sure that the total charge on the ion is correct. The oxidation state of the metallic atom does not change.

Example: A solution of iron (III) nitrate is added to a solution of potassium thiocyanate.

Solution: $Fe^{3+} + SCN^- \rightarrow FeSCN^{2+}$

Oxidation-Reduction Reactions

One element increases in oxidation state while another is reduced in oxidation state.

1. **Two uncombined elements come together**

Combine the two elements to form a compound with reasonable oxidation states for each element.

Example: Magnesium metal is burned in oxygen gas.

Solution: $Mg + O_2 \rightarrow MgO$

2. **A carbon compound combusts with oxygen**

An alkane, alkene, or alkyne is oxidized by oxygen gas to form carbon dioxide and water.

Example: Butane gas ignites in the presence of oxygen gas.

Solution: $C_4H_{10} + O_2 \rightarrow CO_2 + H_2O$

3. **A single reactant decomposes**

   Decomposition usually occurs because an uncommon oxidation state in one element gives way to a more common oxidation state.

   Example: Hydrogen peroxide solution is exposed to bright light.

   Solution: $H_2O_2 \rightarrow O_2 + H_2O$

4. **A solid transition metal is placed in a solution of metallic ions**

   Use the chart of standard reduction potentials. The change with the highest reduction potential is reduced in charge. Voltaic or galvanic cells are an example of this type of reaction.

   Example: Copper metal is placed in a solution of silver nitrate.

   Solution: $Cu + Ag^+ \rightarrow Cu^{++} + Ag$

5. **An electrical current passes through a solution**

   If so, this reaction takes place in an electrolytic cell. Only a limited number of possible reactions are possible.

   Example: An electrical current runs between two electrodes in molten sodium chloride.

   Solution: $NaCl \rightarrow Na + Cl_2$

6. **A solid metal is placed into an acidic solution**

   The metal is oxidized and hydrogen gas is formed. Remember, water can also be an acid; check to be sure the dissociation of water on the reduction potential chart shows a higher tendency to be reduced, such as the case with calcium in water.

   Example: Magnesium metal is placed into a weak solution of hydrochloric acid.

   Solution: $Mg + H^+ \rightarrow Mg^{++} + H_2$

RELATIONSHIPS IN THE PERIODIC TABLE

Atomic Radii

- Moving from left to right across a period, atomic radius decreases.
- Moving down a group, atomic radius increases.
- Cations have smaller radii than their corresponding neutral atoms.
- Anions have larger radii than their corresponding neutral atoms.

Ionization Energy

- Moving from left to right across a period, ionization energy increases.
- Moving down a group, ionization energy decreases.
- More energy is needed for each succeeding ionization.
- Significantly more energy is needed to break a full shell of electrons.
- Elements with low ionization energies are more easily oxidized.

Electron Affinity

- Moving from left to right across a period, electron-affinity energy given off increases.
- Moving down a group, electron-affinity energy given off does not change appreciably.

Electronegativity

- Moving from left to right across a period, electronegativity increases.
- Moving down a group, electronegativity decreases.

Reactivity of the Main Groups

- **Group IA** elements are the most reactive of the metals. The outer "s" electron is loosely held and easily removed—and more easily removed for larger elements. Group IA elements react violently with water and are so reactive that they are never found naturally in an uncombined state. Hydroxides of IA elements are strong bases; increased size of the IA element will increase the base strength of the element's hydroxide.

- **Group IIA** are also very reactive and do not tend to be found uncombined in nature. They react slowly with oxygen to form oxides and with water to form hydroxides.

- **Transition metals** all react very slowly with oxygen or water, and some do not react at all. The metals with lower reactivity can exist in uncombined in nature, and therefore were some of the first metals discovered by ancient civilizations.

- The lightest of the **IIIA** group are non-metallic, while the remaining are all metals. Group IIIA elements are all mildly reactive. For example, aluminum resists combining with oxygen or reacting with water at normal temperatures, but it will react with hydrochloric acid.

- The first element of the **Group IVA** elements (carbon) is a non-metal. It is followed by two metalloids (silicon and germanium), and then by two metals (tin and lead). The first three elements are semi-conductors and fairly reactive.

- **Group VA** consists of two non-metals, two metalloids, and a metal. The more reactive of these elements combine readily with oxygen and some of the more reactive metals.

- **Group VIA** elements, such as oxygen, easily gain electrons. This ability decreases with the larger elements.

- **Group VIIA** elements gain electrons most easily, with the most highly reactive element being the smallest in this group (fluorine).

- **Group VIIIA** elements are the inert gases, and because they have a full outer shell of electrons, they do not combine easily with other elements. For example, the smaller inert gases exist as monoatomic gases. The largest in this group can combine with other elements, but only with highly electronegative elements (such as fluorine and chlorine) in coordinate covalent bonds.

INORGANIC COMPOUNDS

Naming Inorganic Compounds

- Monoatomic cations take the name of the element (*e.g.*, "sodium" for Na^+).

- Cations with multiple possible oxidations states should have the oxidation state listed with a Roman numeral (*e.g.*, Tin (IV) oxide for SnO_2).

- Monoatomic anions take the name of the element, but ends with "ide" (*e.g.*, "fluoride" for F^-).

- Name the cation first, then the anion (*e.g.*, "sodium fluoride").

- Use Greek prefixes for multiple ions. *mono-*, *di-*, *tri-*, *tetra-*, *penta-*, *hexa- hepta-*, *octa-*, etc. *E.g.*, dinitrogen pentoxide for N_2O5.

- Be sure to memorize the common cations and anions in the following charts.

Common Cations

Name	Formula
Hydrogen	H^+
Lithium	Li^+
Sodium	Na^+
Potassium	K^+
Cesium	Cs^+
Beryllium	Be^{2+}
Magnesium	Mg^{2+}
Calcium	Ca^{2+}
Barium	Ba^{2+}
Aluminum	Al^{3+}
Iron II / III	Fe^{2+} / Fe^{3+}
Copper I / II	Cu^+ / Cu^{2+}
Tin II / IV	Sn^{2+} / Sn^{4+}
Lead II / IV	Pb^{2+} / Pb^{4+}
Mercury I / II	Hg^+ / Hg^{2+}
Ammonium	NH^{4+}

Common Anions

Name	Formula
Hydride	H^-
Fluoride	F^-
Chloride	Cl^-
Oxide	O^{2-}
Sulfide	S^{2-}
Nitride	N^{3-}
Phosphide	P^{3-}
Carbonate	$CO_3{}^{2-}$
Bicarbonate	$HCO_3{}^-$
Hypochlorite	ClO^-
Chlorite	$ClO_2{}^-$

(Continued)

Common Anions (Continued)

Name	Formula
Chorate	ClO_3^-
Perchlorate	ClO_4^-
Iodite	IO_2^-
Iodate	IO_3^-
Acetate	$C_2H_3O_2^-$
Permanganate	MnO_4^-
Dichromate	$Cr_2O_7^-$
Chromate	CrO_4^-
Peroxide	O_2^{2-}
Nitrite	NO_2^-
Nitrate	NO_3^-
Sulfite	SO_3^{2-}
Sulfate	SO_4^{2-}
Hydrogen sulfate	HSO_4^-
Hydroxide	OH^-
Cyanide	CN^-
Phosphate	PO_4^{3-}
Hydrogen phosphate	HPO_4^{2-}
Dihydrogen phosphate	H_2PO^{4-}

ORGANIC MOLECULES

- The reactions of some of the functional groups, such as carboxylic acids in acid-base reactions or alkane combustion, are discussed in other sections. (See "Products of Chemical Reactions" earlier in this book.)

Alkanes

Alkane	Formula
Methane	CH_4
Ethane	C_2H_6
Propane	C_3H_8
Butane	C_4H_{10}
Pentane	C_5H_{12}
Hexane	C_6H_{14}
Heptane	C_7H_{16}
Octane	C_8H_{18}
Nonane	C_9H_{20}
Decane	$C_{10}H_{22}$

Alkenes and Alkynes

- **Alkenes** are hydrocarbons that contain a double bond between at least two carbons. Alkenes have the general formula, C_nH_{2n}.

- **Alkynes** are hydrocarbons that contain a triple bond between at least two carbons. Alkynes have the general formula, C_nH_{2n-2}.

- The first covalent bond between two non-metals is a **sigma bond** (σ), where the electrons are paired along the axis between the two atoms.

- Any additional covalent bonds between non-metals are **pi bonds** (π), where the electrons are paired through overlap of p-orbitals above and below the inter-nuclear axis. For example, a double bond consists of one sigma and one pi bond. A triple bond consists of one sigma and two pi bonds.

- Sigma bonds are much stronger than pi bonds, and therefore have much higher bond energies and are more difficult to break.

Organic Functional Groups		
Functional Group	**General formula**	**Example**
Halohydrocarbon	H | R — C — Cl | H	Carbon tetrachloride
Alcohol	H | R — C — OH | H	Ethanol
Ether	H　　　　H |　　　　| R — C — O — C — R |　　　　| H　　　　H	Diethyl ether
Aldehyde	H | R — C — H || O	Formaldehyde
Ketone	H | R — C — R || O	Methyl ethyl ketone
Carboxylic acid	R — C — OH || O	Acetic acid
Ester	H 　　　　　　| R — C — O — C — R ||　　　　| O　　　　H	Diphenyl ester

CHAPTER 10

Experimental Chemistry

CHAPTER 10

EXPERIMENTAL CHEMISTRY

LABORATORY SAFETY

1. When diluting acids, "do it like you oughta; add the acid to the water." Adding water to concentrated acid will quickly cause the water to hydrolyze, and the acid will splatter.

2. When lighting a Bunsen burner, use a match or sparking device; do not use a gas cigarette lighter.

3. If heating a material that may be flammable, set the beaker or test tube that contains the material in a water bath over the flame. This indirect heating assures that the temperature of the material will not be heated too quickly and will never exceed the boiling temperature of water.

4. When heating material, heat it slowly and be sure to point the container away from people. Sudden flashing may quickly push the material out of the container.

5. Assume that all chemicals you work with are toxic. Therefore, never touch your face in the laboratory, and always wash your hands after leaving the laboratory. It is easy to get small amounts of substances on your hands. Report any spills to the teacher.

6. Wear proper protective gear: an apron and eye protection.

7. Know how to use emergency equipment, such as the eye-wash station, shower, fire blanket, and fire extinguisher.

LABORATORY MEASUREMENTS

Measurements and Accuracy

1. Understand the difference between precision and accuracy. Accuracy represents how close a measurement is to the real or accepted value. Measurements are precise when multiple measurements are close to each other. For example, you might make an inaccurate standard solution for an acid-base titration. If the standard solution has a lower concentration than thought, more of it would be required to neutralize the sample, and you will think that the sample has a higher concentration. If you attempt the neutralization several times and get the same answer each time, the results would be precise, but not accurate.

2. Be aware of significant figures in laboratory measurements. You are allowed one uncertain figure in your measurement. For example, in the following buret measurement, the volume is delineated by tenths of a milliliter (mL).

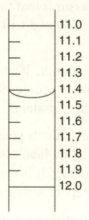

```
11.0
11.1
11.2
11.3
11.4
11.5
11.6
11.7
11.8
11.9
12.0
```

   (a) First, remember to read the bottom of the meniscus when measuring volume in glassware.

   (b) Since one uncertain digit is allowed, you should record 11.45 mL. It might be 11.44 mL; you are uncertain about the +/− 0.01 mL amount, but you are certain that the +/− 0.1 mL aspect of the measurement is between 11.4 and 11.5 mL.

   (c) With regard to significant figures, remember that if they are multiplied or divided together to get some final answer, the number of significant figures in the answer should match the least number of significant figures in the numbers that lead to the answer.

3. When measuring the mass of a substance in a recently heated container, be sure to let the container cool before you put it on the balance. Air currents may decrease the accuracy of the measurement by artificially decreasing the measured mass.

4. When making solutions, mix the solution slowly to be sure that all the solid dissolves. Your results might be affected if the solution is less concentrated than you think.

5. When using a buret for titration, be sure to rinse the buret with the standard solution to be used in the titration. Rinsing with the sample to be titrated will neutralize some of the standard solution before the experiment begins. Rinsing with water will dilute the standard solution before the experiment begins.

6. When you make solutions or use chemicals, you are assuming that they are pure—not a mixture with other chemicals. Therefore, be sure not to inadvertently add impurities. Be sure that a spatula is clean when you put it in a chemical. Do not touch the tip of a pipette to anything accept the solution that it holds.

Interpreting Results

1. First, never expect quantitative measurements to exactly match the expected value. Be sure to calculate the percent accuracy by dividing the difference between the measured and expected values, dividing by the expected value, and multiplying by 100. This figure should be a part of any laboratory report involving quantitative measurements.

2. Analyze the precision of your measurements by making multiple measurements or comparing your value with your classmates. If measurements vary within the +/− accuracy of your measuring devices, they are virtually the same value, even though the number as expressed on your calculator might be different. Use tables or graphs to better summarize multiple data points.

3. As you examine the accuracy of your measurements, consider each step that you undertook to make the measurement. Consider how an inadvertent error in that step could have influenced the measurement, either by increasing or decreasing the value of the measurement relative to the expected value.

4. When graphing a variable as you change another variable, the value that you measure should be on the *y*-axis; this is the *dependent variable*. It has changed as a result of changing the *independent variable*, which should be on the *x*-axis. If such a graph shows an increasing slope, then the dependent variable is *directly proportional* to the independent variable. If such a graph shows a decreasing and rounded slope, then the dependent variable is *inversely proportional* to the independent variable.

EXAMPLE PROBLEMS FROM EXPERIMENTS

1. **Determination of the formula of a compound**

   Example problem:

   A student cuts a 5.30 g magnesium ribbon into small pieces, then began to heat it in a 14.20 g crucible. Over time, the magnesium ignited and burned to form a white, grayish and powdery substance. After the crucible cooled, the student measured the mass of the crucible that contained the powder and found it to be 22.70 g.

   (a) What is the mass of the powder in the crucible?

   22.7 g − 14.20 g = <u>8.50 g powder</u> (5.30 g Mg and 3.20 g O)

   (b) What is the formula of the oxide formed?

   $$\text{Moles of Mg} = 5.30 \text{ g Mg} \times \frac{1 \text{ mol MG}}{24.3 \text{ g MG}} = 0.218 \text{ mol Mg}$$

   $$\text{Moles of O added} = 3.20 \text{ g O} \times \frac{1 \text{ mol O}}{15.99 \text{ g O}} = 0.200 \text{ mol O}$$

   Formula: <u>MgO</u>

   (c) What is the mass percent of oxygen in the compound?

   $$\frac{3.20 \text{ g O}}{8.50 \text{ g MgO}} \times 100\% = \underline{37.6\% \text{ oxygen}}$$

   (d) What is the most likely source of error in the experiment, and how do these results demonstrate that error?

   It is more likely that some of the magnesium did NOT combine with oxygen in an incomplete reaction, and therefore demonstrate slightly fewer moles of oxygen than magnesium, rather than the other way around. The experimental observations demonstrate this type of error (0.200 mol O *vs.* 0.218 mol Mg).

2. **Determination of the percentage of water in a hydrate**

Example problem:

A student obtained 4.23 g of blue-colored hydrate of copper sulfate from the teacher, and then heated the substance in a 16.80 g crucible over moderate heat for several minutes. Over time, the entire substance appeared to turn white, at which time the student removed the crucible from the heat. The student measured the mass of the crucible and substance together and found it to be 19.50 g.

(a) What mass of water was removed from the hydrate?

19.50 g total − 16.80 g crucible = 2.70 g anhydrate

4.23 g hydrate − 2.70 g anhydrate = <u>1.53 g water</u>

(b) How many moles of water were removed from the hydrate?

$$1.53 \text{ g water} \times \frac{1.0 \text{ mole water}}{18.00 \text{ g water}} = \underline{0.085 \text{ moles water}}$$

(c) How many moles of the anhydrous copper sulfate were left in the container after heating?

$$2.70 \text{ g CuSO}_4 \times \frac{1.0 \text{ mole CuSO}_4}{159.61 \text{ g CuSO}_4} = \underline{0.017 \text{ moles CuSO}_4}$$

(d) What is the formula of the hydrate of copper sulfate?

Since 0.085 mol water is five times 0.017 moles copper sulfate, the formula of the hydrate is $CuSO_4 \cdot 5H_2O$.

(e) What is the percentage of water in the hydrate?

$$\frac{5 \times 18.00 \text{ g water}}{249.61 \text{ g CuSO}_4 \cdot 5H_2O} \times 100\% = \underline{36.0\%}$$

(f) What type of error would be created if the student did not heat the copper sulfate long enough to remove all the water?

If less than the entire amount of water were removed, the student would have attributed a greater amount of the original mass to copper sulfate, and less to water. The student would probably end up concluding that the coefficient for water would have been some integer less than 5.

3. **Determination of the molar mass by freezing point depression**

Example problem:

A student is given a 3.80 g sample of an unknown non-electrolyte solid and is asked to find its molar mass by freezing point depression. The student decides to use 40.00 g of benzene as a solvent, which has a melting point of 5.5°C and molal freezing point constant of 5.12°C kg/mol.

(a) What is the significance of the unknown solute being a non-electrolyte? How could the student have figured that property out if she hadn't been told this information?

Freezing point depression is given by the equation $\Delta T = k_f mi$; where ΔT is the amount the freezing point is depressed, k_f is the molal freezing point constant of the solvent, m is the molality of the solvent, and i is the van't Hoff factor—which refers to the degree of ionization of the solute in the solvent. If the unknown solute is a non-electrolyte, then it is not ionized and $i = 1.0$.

(b) Describe how the student should find the freezing point of the unknown/solvent mixture using graphical means.

The student should place the solution in a test tube, immerse the tube in ice, and then measure and record the temperature at regular intervals of time. When the temperature values are plotted vs. time, the slope of the resulting best-fit line will decrease sharply, and then change abruptly. The new melting point of the solution is the temperature that corresponds to the point where the slope of the cooling curve abruptly changes.

(c) The student measures the ΔT to be 3.2 °C. What is the molality of the solution that contains the unknown?

$$m = \frac{\Delta T}{k_f i} = \frac{3.2 \,°C \, mol}{5.12 \,°C \, kg} = 0.625 \, molal$$

(d) How many moles of the unknown were in the test solution?

$$0.025 \text{ moles unknown} = \frac{0.625 \text{ moles unknown}}{\text{kg solvent}} \times$$

$$0.0400 \text{ kg solvent}$$

(e) What is the molar mass of the unknown substance?

$$\text{molar mass} = \frac{\text{grams of unknown}}{\text{moles of unknown}} = \frac{3.80 \text{ g unknown}}{0.025 \text{ mol unknown}} =$$

152 g/mol.

4. **Determination of the molar volume of a gas**

Example problem:

A 0.035 g sample of magnesium ribbon is placed in a stoppered flask that is connected to a eudometer tube, which collects gas by water displacement. When concentrated HCl is placed in the flask with the magnesium, 32.42 mL hydrogen gas is produced in the eudiometer tube.

(a) Write the balanced chemical reaction that occurs when the magnesium is combined with the HCl.

$$\text{Mg } (s) + 2\text{H}^+ (aq) \rightarrow \text{Mg}^{2+} (aq) + \text{H}_2 (g)$$

(b) How many moles of hydrogen gas are expected to be produced from this amount of magnesium in this reaction.

$$.00144 \text{ moles H}_2 (g) = 0.035 \text{ g Mg} \times \frac{1 \text{ mol Mg}}{24.31 \text{ g Mg}} \times$$

$$\frac{1 \text{ mol H}_2 (g)}{1 \text{ mole Mg}}$$

(c) How many moles of water also exists with the collected hydrogen gas in the eudiometer tube? (Vapor pressure of water at 22 °C = 0.030 atm.)

$$n = \frac{PV}{RT} = \frac{0.030 \text{ atm (K mol)}}{0.082 \text{ L atm}} \times \frac{0.03242 \text{ L}}{295 \text{ K}} =$$

4.02×10^{-5} mol water

(d) What is the mole fraction for the water vapor that exists in the eudiometer tube?

$$\text{Mole fraction} = \frac{\text{moles water}}{\text{total moles}} = \frac{\text{pressure water}}{\text{total pressure}} = 0.03$$

(e) What is the percent error between the number of moles of hydrogen gas actually produced at 1.0 atm and 22 °C and the expected yield of hydrogen gas?

$$0.0013 \text{ moles H}_2 = \frac{PV}{RT} = \frac{(1.0 - 0.03) \text{ atm (K mol)}}{0.082 \text{ L atm}} \times \frac{.03242 \text{ L}}{295 \text{ K}}$$

$$\frac{0.00144 - 0.00130}{0.00144} \times 100\% = 9.7\% \text{ error.}$$

5. **Standardization of a solution using a primary standard**

Example problem:

A student wishes to use a known concentration of $KMnO_4$ for an oxidation titration in Experiment #8 below. However, to do this experiment, the student must know exactly the concentration of a stock solution of $KMnO_4$ that is in the storeroom. She carefully mixes a solution of 0.10-molar oxalic acid to react with the $KMnO_4$ in the following reaction:

$$5 H_2C_2O_4 + 2MnO_4^- + 3 H_2SO_4 \rightarrow 2 MnSO_4 + 10 CO_2 + 8 H_2O$$

(a) How does the student know when the oxalic acid uses up all the permanganate ion in the sample?

The permanganate ion is a deep purple color. When the ion is used up, the solution in the beaker will be clear.

(b) Is the oxalic acid oxidized or reduced?

During the reaction the carbon atom in the oxalic acid molecule goes from an oxidation state of +3 to an oxidation state of +4. Since the oxidation state of the carbon increases during the reaction, then the carbon in the oxalic acid is oxidized.

(c) 50 mL of the oxalic acid solution is required to completely react with 100 mL of the permanganate solution. What is the concentration of the permanganate solution?

$$\frac{0.050 \text{ L H}_2C_2O_4 \text{ solution}}{0.100 \text{ L MnO}_4 \text{ solution}} \times \frac{0.10 \text{ moles H}_2C_2O_4}{\text{L H}_2C_2O_4 \text{ solution}} \times$$

$$\frac{2 \text{ moles MnO}_4^-}{5 \text{ moles H}_2C_2O_4} = 0.02 \text{ M MnO}_4^-$$

6. **Determination of a concentration by oxidation-reduction titration**

Example problem:

A student wishes to analyze an ore sample for Fe^{2+} content. To do this she crushes the ore sample and then soaks it in concentrated HCl to dissolve the Fe^{2+}; then she oxidizes the Fe^{2+} ion to Fe^{3+} using a standardized solution of potassium permanganate.

(a) Write the balanced oxidation-reduction reaction that the student will undergo in this titration.

$$5Fe^{2+} + 8H^+ + MnO_4^- \rightarrow 5Fe^{3+} + Mn^{2+} + 4H_2O$$

(b) To analyze the sample, the student creates a standardized solution of potassium permanganate by combining 3.20 g of $KMnO_4$ with enough water to yield 200 mL of solution. What is the molar concentration of the standardized solution?

$$\frac{3.20 \text{ g } KMnO_4}{0.200 \text{ L sln}} \times \frac{1.0 \text{ mole } KMnO_4}{158.04 \text{ g } KMnO_4} = 0.10 \text{ M } KMnO_4$$

(c) How does the student know when she reaches the end point, when all the iron (II) has been converted into iron (III)?

As long as there is iron (II) in the solution that is analyzed, the otherwise dark purple permanganate ion will be used up and turn clear. However, when the iron (II) has been used up—such as at the end point—an additional permanganate ion added will remain unreacted and the sample solution will turn purple.

(d) A 10.00 gram sample of the ore is crushed and titrated with the standardized solution. It takes 6.32 mL of the standardized solution to reach the end point. How many moles of iron (II) are in the sample that was analyzed?

$$0.00632 \text{ L } KMnO_4 \text{ sln} \times \frac{0.10 \text{ mole } KMnO_4}{\text{liters } KMnO_4 \text{ sln}} \times$$

$$\frac{5 \text{ mol } Fe^{2+}}{1 \text{ mole } KMnO_4} = .0032 \text{ moles } Fe^{2+}$$

(e) What is the mass percent of iron (II) in the sample?

$$0.0032 \text{ mol } Fe^{2+} \times \frac{55.85 \text{ g } Fe^{2+}}{1 \text{ mol } Fe^2} \times \frac{1}{10.00 \text{ g total}}$$

$$\times 100\% = 1.8\% \ Fe^{2+}.$$

7. **Determination of mass and mole relationships in a chemical reaction**

This experiment can easily be done separately, or in conjunction with some of the other recommended experiments. For example, the mass-mole questions could easily accompany the first experiment with magnesium.

Example problem:

A student cuts a 2.80 g magnesium ribbon into small pieces, then began to heat it in a 14.20 g crucible. Over time, the magnesium ignited and burned to form a white, grayish and powdery substance. After the crucible cooled, the student measured the mass of the crucible that contained the powder and found it to be 18.73 g.

(a) What is the mass of the powder in the crucible?

18.73 g − 14.20 g = 4.53 g powder (2.80 g Mg and 1.73 g O)

(b) How many moles of magnesium were in the crucible when the reaction began?

$$\text{Moles of Mg} = 2.80 \text{ g Mg} \times \frac{1 \text{ mol Mg}}{24.3 \text{ g Mg}} = 0.115 \text{ mol Mg}$$

(c) Approximately how many moles of oxygen were added to the magnesium during the reaction?

$$\text{Moles of O added} = 1.73 \text{ g O} \times \frac{1 \text{ mol O}}{15.99 \text{ g O}} = 0.108 \text{ mol O}.$$

8. **Determination of the equilibrium constant for a chemical reaction**

Example problem:

A student is given a solution of acetic acid and is asked to find the equilibrium constant for the following equilibrium.

$HC_2H_3O_2 \ (aq) \rightarrow H^+ \ (aq) + C_2H_3O_2^- (aq)$

(a) Describe how the student can use this information and a pH meter to find the K_a of acetic acid.

The student can make a standardized solution of acetic acid, then measure its pH. The pH can tell the student how much the acetic acid has dissociated. From the amount dissociated, the student can determine the concentrations of all species after equilibrium, and then calculate the equilibrium constant using the relationship:

$$K_a = \frac{[H^+][C_2H_3O_2^-]}{[HC_2H_3O_2]}$$

(b) Calculate the initial concentration (before equilibrium is established) of the sample of acetic acid if 2.0 mL of an initial 12.0 M stock solution is diluted to make 400 mL of solution.

$$(2.0 \text{ mL}) \times (12.0 \text{ M}) = (400 \text{ mL}) \times (X)$$

$$\frac{2.0 \text{ mL} \times 12.0 \text{ M}}{400 \text{ mL}} = X = \underline{.06 \text{ M}}$$

(c) After equilibrium is established, the student measures the pH to be 2.98. What is the dissociation constant (K_a) for this reaction?

$$HC_2H_3O_2 (aq) \leftrightarrow H+ (aq) + C_2H_3O_2^- (aq)$$

Equilibrium concentrations: 0.06M $10^{-2.98}$ $10^{-2.98}$

Plugging into the equilibrium expression:

$$K_a = \frac{[H^+][C_2H_3O_2^-]}{[HC_2H_3O_2]} = \frac{(10^{-2.98})^2}{(0.06)} = 1.8 \times 10^{-5}$$

(d) What is a reasonable approximation for the standard free energy, $\Delta G°$, for this reaction?

$$\Delta G° = -RT \ln K = -(8.31 \text{ J}/\text{K mol})(300 \text{ K}) \ln 1.8 \times 10^{-5} = 27,000 \text{ J}.$$

9. Determination of the rate of a reaction and its order

Example problem:

A student runs an experiment that measures the rate of reaction between iodine and acetone at varied concentrations, according to the following reaction:

$$C_3H_6O + I_2 \rightarrow C_3H_6OI + H^+ + I^-$$

The student measures the time it takes for the iodine to disappear when varying the concentrations in the following four trials.

[acetone] (M)	[H$^+$] (M)	[I$_2$] (M)	rate of reaction (M/sec)
0.100	0.100	0.100	2.6×10^{-7}
0.100	0.100	0.200	2.6×10^{-7}
0.100	0.200	0.400	5.2×10^{-6}
0.200	0.200	0.400	1.0×10^{-5}

(a) Determine the order of the reaction with respect to each chemical species tested.

The reaction is first order with respect to acetone and H^+, and zero order with respect to iodine.

(b) What is the rate equation for this reaction?

Rate $= k\,[C_3H_6O]\,[H^+]$

(c) What are the value and units for the rate constant, k?

2.6×10^{-5} L / M sec.

10. **Determination of the enthalpy change associated with a reaction.**

Example problem:

A student wishes to calculate the molar heat of combustion of butane, and is given only a beaker, a thermometer, and a small disposable cigarette lighter.

(a) Design an experiment and determine what needs to be measured in order to find the molar heat of combustion of butane.

The student can use the butane in the lighter to heat water and measure the temperature change in the water. Weighing the lighter before and after the reaction will allow the student to calculate the number of moles of butane used. Multiplying the change in temperature of the water by the mass of the water, by its specific heat (4.18 J/g°C), will tell the student how much heat was absorbed by the water. This type of "coffee cup calorimetry" is crude, but gives a reasonable approximation of an enthalpy of combustion.

(b) The student weighs the lighter to be 8.72 grams before starting. The lighter is found to only weigh 6.88 g after the student held the lighter open for several minutes under the beaker that contained 400 g of water. During this time, the temperature of the water rose by 22.4 °C.

(i) How many moles of butane were used up?

1.84 g butane $\times\ \dfrac{1\ \text{mol butane}}{58.0\ \text{g butane}} =$

0.0317 moles butane used up

(ii) How much heat was absorbed by the water?

$q = \text{mc}\Delta T = 400\ \text{g} \times 4.18\ \text{J/g°C} \times 22.4\ °C = 3.75 \times 10^4\ \text{J}$

(iii) What is the molar heat of combustion for butane that can be calculated from these measurements?

$$\frac{3.75 \times 10^4 \text{ J}}{0.0317 \text{ mol}} = 1.18 \times 10^6 \text{ J/mol} = 1.18 \times 10^3 \text{ kJ/mol}$$

(c) What is the most typical type of error that this student will get in her measurement, and in which direction will this error push the results?

Typically, not all the heat emitted by the reaction will go into raising the temperature of the water; some will be lost to the surroundings. This will result in a measured value that is significantly less kJ/mol than the expected value.

11. Analytical gravimetric determination

Example problem:

A student is given 2.482 grams of a powdered mixture that contains both $MgCl_2$ and $CaCO_3$. She is then to find the mass percent of the magnesium chloride in the mixture. To begin the analysis, the student dissolves the solid in strong acid, and then precipitates one of the compounds with lead nitrate.

(a) What will happen to the calcium carbonate when the acid is added to the mixture?

The acid will dissolve the calcium carbonate; carbon dioxide gas will be emitted, leaving the dissolved calcium in solution.

(b) What will be the precipitate when the lead nitrate is added to the solution. Given the dependence on this reaction, which strong acid should NOT be used in the first step?

The soluble lead ions will combine with the chloride ions to form insoluble lead chloride. For this reason, HCl should NOT be used in the first step. It would add chloride ions and result in an inaccurately high value for the percent of magnesium chloride in the mixture.

(c) When the lead nitrate is added, 5.280 g of precipitate is formed. What is the mass percent of magnesium chloride in the original mixture?

$$5.280 \text{ g PbCl}_2 \times \frac{1 \text{ mol PbCl}_2}{278.1 \text{ g PbCl}_2} \times \frac{2 \text{ mol Cl}}{1 \text{ mol PbCl}_2} \times$$

$$\frac{1 \text{ mol MgCl}_2}{2 \text{ mol Cl}} \times \frac{95.21 \text{ g MgCl}_2}{1 \text{ mol MgCl}_2} = 1.807 \text{ grams MgCl}_2$$

Mass percent = 1.807 g / 2.482 g × 100% = <u>72.8% MgCl$_2$</u>

(d) What would be the maximum volume of gas collected during the first step if the surrounding temperature and pressure were 300 K and 0.92 atm, respectively?

0.675 g of calcium carbonate remain.

$$0.675 \text{ g CaCo}_3 \times \frac{1 \text{ mol CaCO}_3}{100.09 \text{ g CaCO}_3} \times \frac{1 \text{ mol CO}_2}{1 \text{ mol CaCO}_3} =$$

0.0067 mol CO$_2$

$$V = \frac{nRT}{P} = \frac{(0.0067 \text{ mol}) (0.082 \text{ L atm}) (300 \text{ K})}{(0.92 \text{ atm}) \text{ K mol}} =$$

<u>0.179 L CO$_2$</u>.

12. Preparation and properties of buffer solutions

Example problem:

(a) Describe briefly how buffers work. Use specific chemical reactions with the acetic acid/acetate system to show how it minimizes a change in pH when either a strong acid or a strong base is added to the system.

A buffer is a combination of a weak acid and its conjugate base. When this combination exists, any additional protons added to the solution would react with the conjugate base to form water, and little change in pH results.

$$C_2H_3O_2^- + H^+ \rightarrow H_2O$$

Likewise, any additional hydroxide ions added to the solution would react with the weak acid to form water, and little change in the pH results, except with the very slight increase in pH due to the addition of the conjugate base.

$$HC_2H_3O_2 + OH^- \rightarrow H_2O + C_2H_3O_2^-$$

(b) Use the Henderson-Hasselbach equation to describe why the ideal buffer pH for a buffer system equals the pK_a of the weak acid.

The Henderson-Hasselbach equation shows the relationship between the pK_a and the pH, at different concentrations of the weak acid and the conjugate base.

$$pH = pK_a + \log\frac{[A^-]}{[HA]}$$

The maximum buffering ability exists when the buffer system is equally equipped to neutralize added acid or base, or where the molar concentration of the weak acid equals the concentration of its conjugate base, (where the $[A^-]/[HA]$ ratio equals 1.0.) When $[A^-]/[HA] = 1.0$, then the log of that ratio equals zero, and the pH = pK_a of the acid.

(c) Design an experiment to test the buffering ability of a buffer system. Be sure to identify what variable(s) you would keep constant, and what variable(s) you would change.

One way to test buffering ability is to test different buffer systems. Another type of test would be to look at buffering capability when the members of the buffer are at different concentrations. For each of the different systems you would test, you would use an appropriate indicator that would change color when the buffer is exhausted, and then keep adding a measured amount of acid (or base) until the system reaches its end point. A comparison of the amount of acid (or base) added would give a measure of the system's ability to buffer changes in pH.

(d) Given a 350 mL of a 0.20 M solution of acetic acid, what mass of solid sodium acetate would you add to the solution to achieve the maximum buffering ability of the solution?

$$0.350 \text{ L acid soln.} \times \frac{0.2 \text{ mol acid}}{1 \text{ L acid soln}} \times \frac{1 \text{ mol acetate}}{1 \text{ mol acid}} \times$$

$$\frac{82.0 \text{ g NaC}_2\text{H}_3\text{O}_2}{1 \text{ mol acetate}} = 5.74 \text{ g.}$$

13. **Determination of electrochemical series**

Example problem:

(a) In the following figure, identify the following components of a galvanic cell.

   (i)   voltmeter

   (ii) anode

   (iii) cathode

   (iv) salt bridge

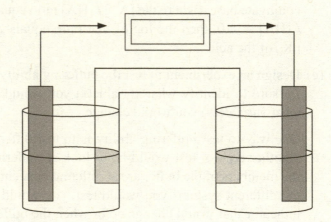

(b) Where does reduction take place?

   At the cathode.

(c) In the figure above, the solution in the beaker on the right is 1.0 M $AgNO_3$, and the electrode in that beaker is solid silver (Ag). Use the chart of standard reduction potentials to calculate the total cell potential when the beaker on the left is filled with each of the following solutions, separately; and the electrode is composed of the corresponding solid metal.

   (i)   1.0 M $Al(NO_3)_3$ (2.46 Volts)

   (ii) 1.0 M $Zn(NO_3)_2$ (1.56 Volts)

   (iii) 1.0 M $Fe(NO_3)_2$ (1.24 Volts)

   (iv) 0.001 M $Cu(NO_3)_2$ (0.44 Volts)

   Based on the differences between the standard reduction potentials, the voltage potentials are seen when these solutions are used to build a voltaic cell along with $AgNo_3$.

14. Measurements using electrochemical cells and electroplating

Example problem:

Two carbon rods, each with a mass of 8.000 g, are put into a 100.0 mL solution of 2.0 M $CuSO_4$. Electrodes from the two poles of a power source are placed at each electrode and the power is turned on for 20 minutes. An ammeter indicates that the current is 2.0 A for the entire time.

(a) What is the reaction at the cathode?

$$Cu^{2+}(aq) + 2\,e^- \rightarrow Cu\,(s)$$

(b) What is the reaction at the anode?

$$2H_2O(l) \rightarrow 4H^+(aq) + O_2\,(g) + 4\,e^-$$

(c) What is occurring to the pH of the solution as the experiment proceeds?

Protons are being formed, so the pH decreases as the experiment proceeds.

(d) How many moles of electrons were delivered to the cathode during the experiment?

$$2.00\ A \times 20.0\ min \times \frac{60\ sec}{min} \times \frac{C}{A\ sec} \times \frac{1\ mole\ e^-}{96{,}486\ C} =$$

$\underline{0.0249\ mol\ e^-}$

(e) What is the predicted new mass of the cathode when it is dried and re-weighed after the experiment?

$$0.0249\ mol\ electrons \times \frac{1\ mole\ copper}{2\ moles\ electrons} \times 63.55\ g\ Cu =$$

$\underline{0.7911\ g\ Cu}$

Final mass electrode = initial mass + mass Cu added = $\underline{8.7911\ g}$.

(f) What is the $[Cu^{2+}]$ of the solution after the experiment is complete?

Moles Cu^{2+} before the reaction = 0.100 L × 2.0 moles = 0.200 moles

Moles Cu^{2+} after the reaction = 0.200 mol − 0.0125 = 0.1875 moles

$[Cu^{2+}]$ = 0.1875 moles / 0.10 L = 1.875 M.

PRACTICE TEST 1

CLEP Chemistry

Also available at the REA Study Center (*www.rea.com/studycenter*)

This practice test is also offered online at the REA Study Center. Since all CLEP exams are computer-based, we recommend that you take the online version of the test to simulate test-day conditions and to receive these added benefits:

- **Timed testing conditions** – helps you gauge how much time you can spend on each question
- **Automatic scoring** – find out how you did on the test, instantly
- **On-screen detailed explanations of answers** – gives you the correct answer and explains why the other answer choices are wrong
- **Diagnostic score reports** – pinpoint where you're strongest and where you need to focus your study

PRACTICE TEST 1

CLEP Chemistry

(Answer sheets appear in the back of the book.)

TIME: 90 Minutes
75 Questions

PART A

DIRECTIONS: Each of the lettered choices below refers to the numbered questions or statements immediately following it. Select the lettered choice that best answers each question and fill in the corresponding oval on the answer sheet. A choice may be used once, more than once, or not at all.

Questions 1–3

    (A) H_2
    (B) $KMnO_4$
    (C) MgO
    (D) KCl
    (E) Fe_2O_3

1. This substance is a very strong oxidizing agent.

2. The metal in this substance has an oxidation number of $+2$.

3. The oxidation potential of this substance is zero.

Questions 4–6

    (A) O
    (B) Cl
    (C) Fr
    (D) N
    (E) Ar

4. The least electronegative element.

5. All of the electrons in the outer p-orbitals of this element have parallel spin.

6. When bonded with itself as a diatomic gas, the resulting compound will have two pi bonds.

Questions 7–9

    (A) $2H_2O_2 \ (l) \rightarrow O_2 \ (g) + 2H_2O$
    (B) $HC_2H_3O_2 + OH^- \rightarrow C_2H_3O_2^- + H_2O$
    (C) $N_2 \ (g) + 3 \ H_2 \ (g) \rightarrow 2 \ NH_3 \ (g)$
    (D) $HCl + NaOH \rightarrow H_2O + NaCl$
    (E) $Al(OH)_3 + 3 \ H^+ \rightarrow Al^{3+} + 3 \ H_2O$

7. Decreasing the pH will cause this reaction to go to the right.

8. Increasing the pressure will cause this reaction to go to the right.

9. Shows a decomposition reaction.

For questions 10–13, identify the answer that corresponds with the correct functional group.

(A)

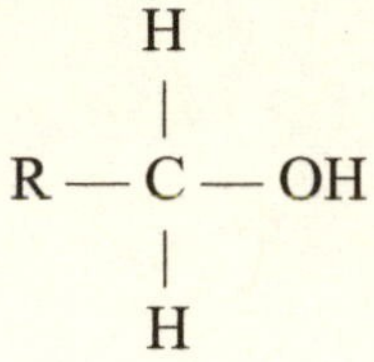

$$\text{R} - \overset{\displaystyle H}{\underset{\displaystyle H}{\overset{|}{\underset{|}{C}}}} - \text{OH}$$

(B) $\text{R} - \overset{\displaystyle C}{\underset{\displaystyle O}{\overset{||}{}}} - \text{H}$

(C)

$$\text{R} - \overset{\displaystyle C}{\underset{\displaystyle O}{\overset{||}{}}} - \text{O} - \overset{\displaystyle H}{\underset{\displaystyle H}{\overset{|}{\underset{|}{C}}}} - \text{R}$$

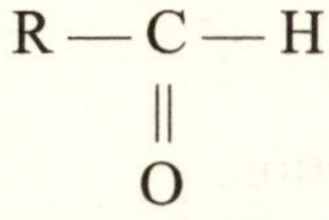

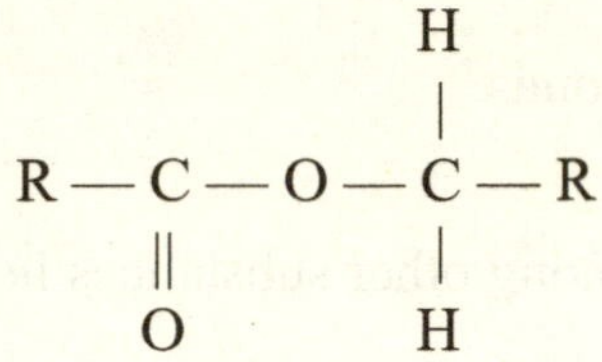

(D) $\text{R} - \overset{\displaystyle C}{\underset{\displaystyle O}{\overset{||}{}}} - \text{R}$

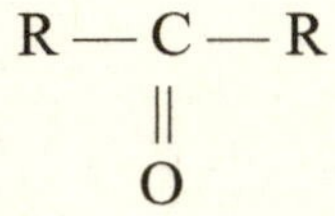

(E) $\text{R} - \overset{\displaystyle C}{\underset{\displaystyle O}{\overset{||}{}}} - \text{OH}$

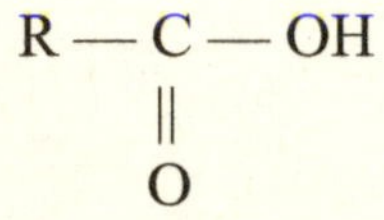

10. Ketone

11. Alcohol

12. Ester

13. Aldehyde

Questions 14–17

    (A) N_2
    (B) K
    (C) H_2
    (D) Br_2
    (E) Xe

14. This substance is a liquid at room temperature.

15. This substance has the weakest intermolecular attraction.

16. The molecules in this gas contain triple bonds.

17. This gas has the lowest rate of effusion among other substances listed when they are at room temperature.

Questions 18–20

    (A) KCl
    (B) NaF
    (C) Cu
    (D) SiC
    (E) $HC_2H_3O_2$

18. Atoms are held together with network covalent attraction.

19. Atoms are held together with covalent attraction.

20. Atoms are held together with metallic attraction.

Questions 21–23

(A) $C_4H_{10} + O_2 \rightarrow CO_2 + H_2O$
(B) $HC_2H_3O_2 + OH^- \rightarrow C_2H_3O_2^- + H_2O$
(C) $Pb^{++} + 2\,I^- \rightarrow PbI_2$
(D) $BF_3 + NH_3 \rightarrow BF_3NH_3$
(E) $Fe^{3+} + SCN^- \rightarrow FeSCN^{2+}$

21. This reaction is a Lewis acid-base reaction.

22. This reaction is a precipitation reaction.

23. This reaction is an oxidation-reduction reaction.

Questions 24–25

(A) Rutherford-fired alpha particles at gold foil
(B) Bohr observed that hydrogen had a unique spectrum when electrons were excited
(C) Thompson observed the deflection of particles in a cathode ray tube
(D) Schrödinger calculated the wave function of electrons
(E) Dalton measured the mass of reactants and products in reactions

24. Determined that electrons had a negative charge

25. Determined that electrons existed a definite and discreet distance from the nucleus

PART B

26. Which of the following solid crystals has, on average, one atom per cubic unit cell?

    (A) Face-centered
    (B) Body-centered
    (C) Rhombic
    (D) All cubic crystals
    (E) Simple cubic crystals only

27. Which of the following is the normal vapor pressure of water at 373 K?

    (A) 0.001 atm
    (B) 0.10 atm
    (C) 1.0 atm
    (D) 10.0 atm
    (E) Cannot be determined with this information

28. Use the following chart to determine which set of letters represents an increase in entropy.

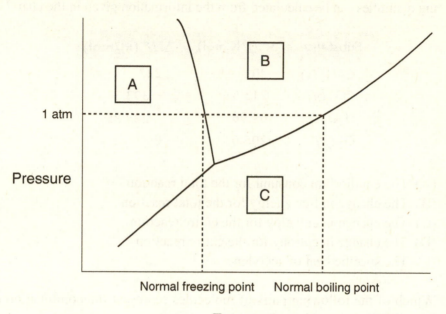

(A) A, B, C
(B) C, B, A
(C) B, C, A
(D) B, A, C
(E) Cannot be determined from this information

29. Henry's law pertains to which of the following phenomena?

(A) The relationship between temperature and pressure, at constant pressure
(B) The relationship between pressure and volume, at constant temperature
(C) The relationship between free energy and the equilibrium constant
(D) The relationship between concentration and pH
(E) The relationship between the amount of gas dissolved in a solution and the partial pressure of the gas above the solution

30. The following chart contains thermodynamic information about reactants and products in the combustion reaction for acetylene, C_2H_2. Which of the following quantities can be calculated from the information given in the chart?

Substance	$S°$ (J/K mol)	$\Delta H_f°$ (kJ/mol)
C_2H_2 (s)	201	227
CO_2 (g)	213.6	-393.5
H_2O (l)	69.96	-285.85
O_2 (g)	205.0	0

(A) The equilibrium constant for the total reaction
(B) The change in free energy for the total reaction
(C) The change in enthalpy for the entire reaction
(D) The change in entropy for the entire reaction
(E) The specific heat of acetylene

31. Which of the following pairs of molecules represent the combination of a Brønsted-Lowry acid and its conjugate base?

(A) Cu/Cu^{++}
(B) N_2/NH_3
(C) $HC_2H_3O_2/C_2H_3O_2^-$
(D) $Al(OH)_3/[Al(OH)_4]^-$
(E) PbI_2/Pb^{2+}

32. Which of the following sets of quantum numbers (listed in order of n, l, m_l, m_s) describe the highest energy valence electron of nitrogen in its ground state?

(A) $2, 0, 0, +\frac{1}{2}$
(B) $2, 1, 1, -\frac{1}{2}$
(C) $2, 1, 1, +\frac{1}{2}$
(D) $2, 1, -1, -\frac{1}{2}$
(E) $2, 1, -1, +\frac{1}{2}$

33. Given the information in this chart, which of the following answers best represents the enthalpy of combustion for hydrogen gas, in kJ/mol?

Substance	ΔH_f° (kJ/mol)
H_2 (s)	0
CO_2 (g)	-393.5
H_2O (l)	-285.85
O_2 (g)	0

(A) 0
(B) -393
(C) -107
(D) -285
(E) $+107$

34. In the following reaction, how would the equilibrium constant for the listed reaction be related to the acid ionization constant, K_a, for acetic acid?

$$HC_2H_3O_2 + OH^- \rightarrow H_2O + C_2H_3O_2^-$$

(A) K_a/K_w
(B) K_w/K_a
(C) $1/K_a$
(D) K_a
(E) K_aK_b

35. What is the molar mass of C_2H_4O, in g/mol?

(A) 12
(B) 22
(C) 32
(D) 44
(E) 60

36. What mass (in grams) of sodium hydroxide, NaOH, would be needed to create 2.0 liters of a 0.40-molar solution of NaOH?

(A) 0.8
(B) 8.0
(C) 16
(D) 19.2
(E) 32

37. What is the approximate percent composition of oxygen in magnesium oxide, MgO?

    (A) 12%
    (B) 20%
    (C) 32%
    (D) 40%
    (E) 80%

38. What would be the most likely rate law for the mechanism below?

    Step 1: $A + B \rightarrow I$ fast equilibrium
    Step 2: $\underline{C + I \rightarrow D}$ slow step
    Total reaction: $A + B + C \rightarrow D$

    (A) Rate $= k\,[A]\,[B]$
    (B) Rate $= k\,[A]\,[B]\,[C]$
    (C) Rate $= k\,[C]$
    (D) Rate $= k\,[C]\,[I]$
    (E) Cannot be determined from the information given

39. The K_{sp} of $PbCrO_4$ is 1.0×10^{-16}. What is the molar solubility of $PbCrO_4$ in a solution with pH of 4?

    (A) 1.0×10^{-4}
    (B) 1.0×10^{-8}
    (C) 1.0×10^{-16}
    (D) 1.0×10^{-20}
    (E) 1.0×10^{-22}

40. The following chart depicts the ionization constants (K_a) for a number of weak acids. If you had a 1.0-molar solution of each acid, which would have the highest pH?

Acid	K_a
HSO_2^-	1.2×10^{-2}
HNO_2	4.0×10^{-4}
HF	7.2×10^{-4}
HOCl	3.5×10^{-8}
HCN	6.2×10^{-10}

(A) HSO_2^-
(B) HNO_2
(C) HF
(D) HOCl
(E) HCN

41. What is the ratio of the rate of effusion of hydrogen gas (molar mass = 4.0 g/mol) to the rate of effusion of oxygen gas (molar mass = 32.0 g/mol) at the same temperature and pressure?

(A) 2
(B) 4
(C) 0.125
(D) 8.0
(E) 16.0

42. Which of the following aqueous solutions demonstrates the highest boiling point?

(A) 1.0 M MgF_2
(B) 1.0 M NaCl
(C) 1.0 M HCl
(D) 1.0 M glucose
(E) 2.0 M KNO_3

43. The molar solubility of strontium fluoride, MgF_2, is 1×10^{-3} in pure water. What is the K_{sp} for MgF_2?

    (A) 4×10^{-3}
    (B) 4×10^{-6}
    (C) 4×10^{-9}
    (D) 2×10^{-3}
    (E) 1×10^{-3}

44. Which of the following is the least polar molecule?

    (A) H_2
    (B) H_2O
    (C) H_2S
    (D) C_2H_2
    (E) NaH

45. A set of kinetic experiments for the decomposition of a molecule, A, were recorded in the data table below. What is the order of the reaction with respect to A?

Trial	[A]	Initial rate of formation of C (M/s)
1	0.01	0.002
2	0.03	0.006

    (A) Zero order
    (B) First order
    (C) Second order
    (D) Third order
    (E) Cannot be determined by the information given

46. A sample of calcium carbonate (molar mass = 100 g) is analyzed. Assuming that all the calcium is combined with carbonate, what is the approximate percent of calcium in the sample?

    (A) 20%
    (B) 40%
    (C) 60%
    (D) 75%
    (E) 80%

47. In the following reaction, which species is the reducing agent?

$$Cd + Cu^{2+} \rightarrow Cd^{2+} + Cu$$

(A) Cd
(B) Cu^{2+}
(C) Cu
(D) Cd^{2+}
(E) There is more than one reducing agent

48. What is the $[H^+]$ of a 1.0 M solution of phenol, which has a K_a of 4×10^{-10}?

(A) 1.6×10^{-10}
(B) 2.0×10^{-5}
(C) 4.0×10^{-5}
(D) 2.0×10^{-6}
(E) 4.0×10^{-6}

49. What mass of acetic acid will be consumed by 10.0 grams of NaOH in the reaction below?

$$HC_2H_3O_2 + NaOH \rightarrow H_2O + NaC_2H_3O_2$$

(A) 1.5 grams
(B) 15 grams
(C) 0.66 grams
(D) 6.6 grams
(E) 66 grams

50. What mass of chlorine gas, Cl_2, could be contained in a 3.0-liter flask at 2.0 atm and 400 K?

(A) 0.077 g
(B) 1.0 g
(C) 6.5 g
(D) 13.0 g
(E) 8.5 g

51. For the following reaction, calculate an approximate value for the change in free energy, and determine whether or not the reaction is spontaneous under standard conditions.

$$2\ HCN\ (g) + 4.5\ O_2\ (g) \rightarrow H_2O\ (l) + 2\ CO_2\ (g) + 2\ NO_2\ (g)$$

Compound	ΔG_f° (kJ/mol)
NO_2	52
HCN	125
CO_2	−394
H_2O	−237
O_2	0

(A) −230 kJ/mol, yes
(B) +2300 kJ/mol, yes
(C) −1171 kJ/mol, yes
(D) +230 kJ/mol, no
(E) +2300 kJ/mol, no

52. Calculate the specific rate constant for the reaction $A + B \rightarrow C$ that is represented by the kinetic data in the chart below.

Trial	[A]	[B]	Initial rate of formation of C (M/s)
1	0.10	0.40	0.02
2	0.10	0.80	0.04
3	0.20	0.80	0.16

(A) 2.0 mol²/L² s
(B) 0.1 L²/mol² s
(C) 4.0 mol/L s
(D) 2.5 mol²/L² s
(E) 0.5 L²/mol² s

53. In which of the following solutions will iron sulfide, FeS, be most soluble?

    (A) Pure water
    (B) 2.0 M NaOH
    (C) 1.0 M NaOH
    (D) 2.0 M H_2SO_4
    (E) 2.0 M HCl

54. Which of the following molecules or ions can act as either an acid or a base?

    (A) OH^-
    (B) H_3O^+
    (C) $C_2H_3O_2^-$
    (D) $H_2PO_4^-$
    (E) NH_4^+

55. What is the pressure at Point A on the following phase diagram?

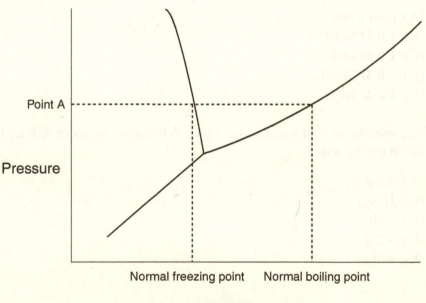

    (A) 1.0 atm
    (B) 1.8 atm
    (C) 2.0 atm
    (D) 5.0 atm
    (E) 10.0 atm

56. Use the information from the following table to calculate the approximate standard change of enthalpy in kJ/mol for the combustion of coal.

Substance	ΔH_f° (kJ/mol)
C (s)	0.0
CO_2 (g)	−393.5
H_2O (l)	−285.85
O_2 (g)	0.0

(A) 0
(B) +286
(C) −698
(D) −286
(E) −394

57. Which of the following aqueous solutions has the lowest melting point?

(A) pure water
(B) 1.0 M NaCl
(C) 1.0 M sucrose
(D) 1.0 M $MgCl_2$
(E) 1.0 M HCl

58. The specific heat of water is 4.2 J/g°C. What mass of water will be heated by 10.0°C by 840 J?

(A) 0.5 g
(B) 10.0 g
(C) 20.0 g
(D) 4.2 g
(E) 840 g

59. Based on the standard reduction potentials listed below, which of the following is the strongest reducing agent?

Fe^{2+}/Fe	-0.44 V
Pb^{2+}/Pb	-0.13 V
Ag^+/Ag	$+0.80$ V

(A) Fe
(B) Fe^{2+}
(C) Pb
(D) Ag
(E) Ag^+

60. A hydrocarbon that contains only carbon and hydrogen is composed of 75% carbon. What is the empirical formula for the compound?

(A) CH_2
(B) CH_3
(C) CH_4
(D) C_2H_5
(E) C_2H_7

61. What is the molar mass of the molecule, NH_3?

(A) 4.0 g/mol
(B) 7.0 g/mol
(C) 10.0 g/mol
(D) 17.0 g/mol
(E) 20.0 g/mol

62. Which of the following has a van't Hoff factor approximately equal to 1.2?

(A) NaCl
(B) $MgCl_2$
(C) $HC_2H_3O_2$
(D) H_2SO_4
(E) glucose

63. What mass of HCl is needed to dilute to 1.0 liter to create a solution with pH of 1.0?

    (A) 0.10 g
    (B) 35 g
    (C) 0.35 g
    (D) 3.5 g
    (E) 7.0 g

64. A student heated a 120-gram sample of a hydrated salt. When it was heated to the point when all the water was removed, the remaining powder weighed 30 grams. What is the ratio of the mass of salt to the mass of water in the hydrate?

    (A) 2:1
    (B) 3:1
    (C) 4:1
    (D) 1:3
    (E) 1:4

65. Which of the following variables is NOT needed to find the molar mass of a nonelectrolyte through freezing-point depression?

    (A) Molal freezing point constant of the solvent
    (B) Mass of the unknown solute
    (C) Mass of the solvent
    (D) The freezing point of the pure unknown solute
    (E) The freezing point of the pure solvent

66. Two carbon rods are put into a 100.0 mL solution of 2.0 M $CuSO_4$. The electrodes are connected to a power source and turned on. Which of the following needs to be measured in order to calculate the mass of metal that precipitates out of solution?

    (A) The current passing to the electrodes and the length of time the current flows
    (B) Only the amount of current passing through the electrodes
    (C) The voltage difference across the electrodes
    (D) The molar concentration of the dissolved ion in solution
    (E) Both voltage and current

67. What is the molar concentration of a 0.50-liter sample of a monoprotic acid if it were completely neutralized by 20 mL of 0.1-molar NaOH solution?

(A) 0.001 M
(B) 0.004 M
(C) 0.05 M
(D) 0.10 M
(E) 1.0 M

68. The kinetics of a reaction in which two reactants, A and B, react together to form a product, C, is studied, and the results are recorded in the chart below. What is the order of the reaction with respect to A and B, respectively?

Trial	[A]	[B]	Initial rate of formation of C (M/s)
1	0.20	0.40	0.001
2	0.20	0.80	0.001
3	0.40	0.80	0.004

(A) First order with respect to A; first order with respect to B
(B) First order with respect to A; zero order with respect to B
(C) Zero order with respect to A; first order with respect to B
(D) Zero order with respect to A; second order with respect to B
(E) Second order with respect to A; zero order with respect to B

69. Molten NaCl was electrolyzed with a constant current of 2.0 amperes for 120 seconds. What number of coulombs passed through the cell during this time?

(A) 2 C
(B) 24 C
(C) 60 C
(D) 120 C
(E) 240 C

70. Which of the following demonstrates pi bonding?

    (A) OH^-
    (B) H^+
    (C) C_2H_2
    (D) H_2S
    (E) KCl

71. In the following reaction, what is the change in oxidation state of aluminum?

$$Al(OH)_3 + 3 H^+ \rightarrow Al^{3+} + 3 H_2O$$

    (A) 4
    (B) 3
    (C) 2
    (D) 1
    (E) 0

72. Which of the following atoms demonstrates the greatest increase between the third and fourth ionization energies?

    (A) He
    (B) Li
    (C) K
    (D) Mg
    (E) Al

73. Which of the following is the strongest acid?

    (A) $HClO_2$
    (B) $HClO_3$
    (C) $HClO_4$
    (D) $HBrO_4$
    (E) HIO_4

74. As you go down the periodic table and to the left, which of the following traits increases?

    (A) Atomic radius
    (B) Electronegativity
    (C) Electron affinity
    (D) Ionization energy
    (E) Acidity of the oxides

75. What is the most likely explanation for the fact that a sample of solid nickel is attracted into a magnetic field, but a sample of solid zinc chloride is not?

    (A) There are unpaired outer electrons in nickel.
    (B) There is some iron mixed in with the nickel.
    (C) There are unpaired outer electrons in zinc.
    (D) The presence of chlorine keeps zinc from being attracted to the magnet.
    (E) None of the above represents a reasonable explanation for this observation.

PRACTICE TEST 1

Answer Key

1.	(B)	26.	(E)	51.	(C)
2.	(C)	27.	(C)	52.	(E)
3.	(A)	28.	(A)	53.	(D)
4.	(C)	29.	(E)	54.	(D)
5.	(D)	30.	(E)	55.	(A)
6.	(D)	31.	(C)	56.	(E)
7.	(E)	32.	(C)	57.	(D)
8.	(C)	33.	(D)	58.	(C)
9.	(A)	34.	(A)	59.	(A)
10.	(D)	35.	(D)	60.	(C)
11.	(A)	36.	(E)	61.	(D)
12.	(C)	37.	(D)	62.	(C)
13.	(B)	38.	(B)	63.	(D)
14.	(D)	39.	(B)	64.	(B)
15.	(C)	40.	(E)	65.	(D)
16.	(A)	41.	(B)	66.	(A)
17.	(E)	42.	(A)	67.	(B)
18.	(D)	43.	(C)	68.	(E)
19.	(E)	44.	(A)	69.	(E)
20.	(C)	45.	(B)	70.	(C)
21.	(D)	46.	(B)	71.	(E)
22.	(C)	47.	(A)	72.	(E)
23.	(A)	48.	(B)	73.	(C)
24.	(C)	49.	(B)	74.	(A)
25.	(B)	50.	(D)	75.	(A)

PRACTICE TEST 1

Detailed Explanations of Answers

1. **(B)**   The manganese ion in potassium permanganate has a +5 oxidation state, and is therefore readily reduced. When it is reduced, it forces another species to be oxidized. Therefore, the permanganate ion is a very strong oxidizing agent.

2. **(C)**   The magnesium atom that is combined with oxygen has a +2 oxidation state. Oxygen carries a −2 oxidation state, and the sum of the oxidation states of oxygen and magnesium must sum to equal the charge on the compound, which is zero.

3. **(A)**   Hydrogen gas by definition, since it is a standard, has an oxidation potential of zero.

4. **(C)**   Francium, Fr, is the least electronegative element on the periodic table. In fact, it is also the element that most easily gives off an electron, since its first ionization energy is lower than any other element.

5. **(D)**   Nitrogen's outer p-electrons all have a parallel spin, since nitrogen has three outer electrons that distribute themselves evenly among its three p-orbitals.

6. **(D)**   In a double bond, the first of the two bonds is a sigma bond that exists along the internuclear axis. The second bond is a pi bond that occurs from the sideways overlap of unhybridized p-orbitals. The bond between two nitrogen atoms is a triple bond that contains one sigma bond and two pi bonds.

7. **(E)**   By decreasing the pH, the concentration of protons will increase. As a result, the equilibrium will shift to the right in order to use up the protons.

8. **(C)**   This is an application of Le Chatelier's principle, where a reaction will shift to minimize change. As a result of increasing the pressure, the reaction equilibrium will shift to minimize the pressure—the move in the direction that produces the least number of gas moles.

9. **(A)** The formation of oxygen and water from hydrogen peroxide is an example of a decomposition reaction, where one molecule splits into more than one molecule.

10. **(D)** A ketone group has a double bond between oxygen and a carbon atom that is imbedded within a carbon chain.

11. **(A)** An alcohol group has a hydroxide group (OH) with the second bond from oxygen going to a carbon atom.

12. **(C)** An ester group is an ether with an additional double bond to an oxygen from an adjacent carbon.

13. **(B)** An aldehyde group has a double bond between an oxygen atom and a carbon that is one end of a carbon chain, or not otherwise bonded to any other carbons.

14. **(D)** Bromine is the only species listed here that is a liquid at room temperature. Indeed, it is the only diatomic halogen that is a liquid. Those smaller than bromine are gases; the one larger than bromine is a solid.

15. **(C)** Hydrogen gas has the weakest intermolecular attraction of those species listed here. Its covalent bond is not only entirely nonpolar, which minimizes polar intermolecular attractions, but it is also the smallest molecule, which minimizes dispersion attractive forces.

16. **(A)** Nitrogen gas is the only species listed that has triple bonds, which occurs because nitrogen has five electrons in its outer shell, three of which are available for bonding in its p-orbitals.

17. **(E)** The largest gas atom or molecule will have the slowest rate of effusion. Xenon is the largest gas particle listed here.

18. **(D)** SiC is held together with network covalent bonds, which creates a crystal entirely from covalent bonds and confers unusual strength (and a high melting point) upon the crystal.

19. **(E)** Acetic acid is a molecular compound that is held together with covalent bonds, where electrons are shared between atoms so that both atoms in the bond end up having a full octet of electrons.

20. **(C)** As a metal, copper is held together by metallic bonds, which involve delocalized d-orbital electrons.

21. **(D)** A Lewis acid is an electron pair acceptor. In the reactions listed here, boron trifluoride seeks a pair of electrons to satisfy its octet and so reacts with the Lewis base (electron pair donor), ammonia.

22. **(C)** According to the solubility rules, the combination of lead ions with iodide ions will combine to form an insoluble product.

23. **(A)** The combustion of butane is the only reaction listed here in which the oxidation state of atoms changes during the reaction. In this case, the carbon is oxidized, and the oxygen is reduced.

24. **(C)** Thompson's experiment deflected electrons with a cathode ray tube, which only deflects negatively charged particles.

25. **(B)** Bohr's observation of the hydrogen spectrum established that there were specific distances from the nucleus where electrons existed. When excited, the electrons would move to radii further from the nucleus, and then emit a photon of a characteristic wavelength when returning to ground state. Since photons of the same energy and wavelength were constantly emitted, the electrons were constantly moving between places that represented discrete distances from the nucleus.

26. **(E)** A simple cubic crystal is the only unit cell mentioned that has, on average, 1 atom per unit cell. A simple cubic unit cell has 1 atom in each of the 8 corners of the unit cell, but only 1/8 of each of those atoms is ascribed to that particular unit cell. A face-centered unit cell has 4 atoms per unit cell; while a body-centered unit cell has 2 atoms per unit cell.

27. **(C)** 373 K is the normal boiling point of water. By definition, the vapor pressure of any substance at its *normal* boiling point is 1.0 atm. The vapor pressure at the boiling point would decrease at high altitudes, but it would not be called the *normal* boiling point.

28. **(A)** The first set of letters demonstrates a substance moving from solid, to liquid, and then to gas. A substance as a solid has less entropy than it does as a liquid and much less entropy than it does as a gas. Entropy is a state of disorder, or randomness, and as a substance is heated, its increased thermal motion gives it greater disorder, or entropy.

29. **(E)** Henry's law pertains to the relationship between the amount of gas dissolved in a solution and the partial pressure of the gas above the solution.

30. **(E)** All of the information given is thermodynamic data, which cannot be used to calculate the specific heat of a substance. The equilibrium constant

can be calculated from the standard change in free energy, which can be calculated from the heat and entropy of reaction, which in turn can be calculated directly from the information on the chart.

$$\Delta G° = -RT \ln K = \Delta H° - T\Delta S°$$

$$\Delta S° = \Sigma S° \text{ (products)} - \Sigma S° \text{ (reactants)}$$

$$\Delta H° = \Sigma H°_f \text{ (products)} - \Sigma H°_f \text{ (reactants)}$$

31. **(C)** Conjugate acid/base pairs are different from one another by only a proton. Acetic acid has a formula of $HC_2H_3O_2$; to remove a proton from it would yield the formula, $C_2H_3O_2^-$.

32. **(C)** Since nitrogen is in the second row, its highest energy electron is at $n = 2$, which is the first number. The second number signifies that its outer electron is in a p-orbital. The third number indicates the third p-orbital to receive an electron, since nitrogen is the third element in the p-block in the periodic table. The last number is the magnetic spin quantum number and signifies that the highest energy electron is the only electron in the orbital.

33. **(D)** The enthalpy of combustion for hydrogen, upon close inspection, is actually the same as the heat of formation for water.

$$\Delta H°_{reaction} = \Sigma H°_f \text{ (products)} - \Sigma H°_f \text{ (reactants)}$$

$$\Delta H°_{reaction} = -285.85 \text{ kJ/mol}$$

34. **(A)** The equilibrium constant for the listed reaction is $1/K_b$, which, since $K_w = K_a \times K_b$, then equals K_a /K_w.

35. **(D)** The molar mass is found by adding the atomic masses of each atom in the formula. $(2 \times 12.011) + (4 \times 1.008) + (16) = 44$ g/mol.

36. **(E)**

$$32 \text{ g NaOH} = 2.0 \text{ L solution} \times \frac{.40 \text{ mol NaOH}}{\text{L NaOH solution}} \times \frac{40 \text{ g NaOH}}{1 \text{ mol NaOH}}$$

37. **(D)** Percent composition is found by dividing the mass contributed by one element by the total molar mass of the compound and then multiplying by 100%.

$$40\% \text{ O} = \frac{16 \text{ g O}}{40 \text{ g MgO}} \times 100\%$$

38. **(B)** The rate-determining step depends on the intermediate, I, which can be expressed in terms of A and B using the prior fast-equilibrium step. Thus, in this mechanism, the rate is not proportional to just I and C, but A, B, and C.

39. **(B)** The molar solubility can be found by equating the solubility product constant to the product of the molar concentrations of the dissolved ions. Therefore, in this case, the molar solubility is simply the square root of the solubility product constant of $PbCrO_4$.

$$x^2 = 1.0 \times 10^{-16}$$

$$x = 1.0 \times 10^{-8}$$

40. **(E)** The weakest acid, with the lowest K_a, would have the highest pH at any given concentration. HCN is by far the weakest acid on the list, with the smallest ionization constant.

41. **(B)** Graham's law of effusion can be used to calculate how much faster hydrogen gas moves than oxygen gas.

$$\frac{M_{oxygen}}{M_{hydrogen}} = \frac{r_{hydrogen}^2}{r_{oxygen}^2} = \frac{32}{2} = 16$$

$(16)^{1/2} = 4$; hydrogen is 4X faster than oxygen.

42. **(A)** MgF_2 has the highest boiling point because the ionic character of MgF_2 is the strongest.

43. **(C)** The relationship between the solubility product constant and molar solubility for a compound that produces 3 moles of ions for every mole of solid dissolved is as follows:

$$K_{sp} = 4x^3 = 4(1 \times 10^{-3})^3 = 4 \times 10^{-9}$$

44. **(A)** Hydrogen gas is the least polar because it is bonded to itself as a diatomic molecule. Therefore, there is no difference in electronegativity between the two atoms in the bond, and it is entirely nonpolar.

45. **(B)** Since the rate of decomposition increases by a factor of 3 when the concentration of A is increased by a factor of 3, then it is a linear relationship between A and the rate, and it is first order with respect to A.

46. **(B)** Percent composition is found by dividing the mass contributed by one element by the total molar mass of the compound and then multiplying by 100%.

$$40\%C = \frac{40 \text{ g CA}}{100 \text{ g CaCO}_3} \times 100\%$$

47. **(A)** Cadmium is being oxidized in this reaction, which means that the other species are being reduced. Therefore, cadmium is the reducing agent.

48. **(B)** The proton concentration in moles per liter is simply the square root of the ionization constant for a one molar solution of the acid.

$$K_a = \frac{x^2}{(1.0)} = 4.0 \times 10^{-10}$$

$$x^2 = 4.0 \times 10^{-10}$$

$$x = [H^+] = 2.0 \times 10^{-5}$$

49. **(B)**

$$15 \text{ g acid} = 10.0 \text{ g NaOH} \times \frac{1 \text{ mol NaOH}}{40 \text{ g NaOH}} \times \frac{1 \text{ mol acid}}{1 \text{ mol NaOH}} \times \frac{60 \text{ g acid}}{1 \text{ mol acid}}$$

50. **(D)**

$$\text{mass} = \frac{PV(\text{molar mass})}{RT} = \frac{2.0 \text{ atm} \times \text{K mol} \times 3.0 \text{ L} \times 70.90 \text{ g}}{(0.082 \text{ L atm}) (400 \text{ K})} = 13 \text{ grams}$$

51. **(C)** The reaction is spontaneous because the change in free energy is less than zero.

$$\Delta G^\circ_{\text{reaction}} = \Sigma \Delta G^\circ_f \text{ (products)} - \Sigma \Delta G^\circ_f \text{ (reactants)}$$

$$\Delta G^\circ_{\text{reaction}} = [(1 \times -237) + (2 \times -394) + (2 \times 52)] - (2 \times 125)$$

$$\Delta G^\circ_{\text{reaction}} = -1,171 \text{ kJ}$$

52. **(E)** Since the rate equation is third order (rate $= k [A]^2 [B]$), then solving for the specific rate constant would yield the following:

$$k = \frac{\text{rate}}{[A]^2[B]} = \frac{.02 \, M/s}{-.04 \, M^3} = \frac{0.5 \, L^2}{\text{mol}^2 \, s}; \text{ since } M = \frac{\text{moles}}{L}$$

53. **(D)** The sulfide ion is used up and removed from the reaction by the acid, because hydrogen sulfide gas is produced. With the removal of the sulfide ion, mass action draws more and more iron sulfide into solution. In this way, the solubility of iron sulfide is maximized in an acidic solution. Option D is the answer that involves the highest concentration of acidic solution and will therefore allow the most iron sulfide to dissolve.

54. **(D)** Only the dihydrogen phosphate ion is able to either donate a proton (in the presence of a base) or accept a proton (in the presence of a very strong acid). The hydroxide ion (option A) cannot donate a proton. The hydronium ion (option B) cannot accept a further proton. The acetate ion (option C) cannot donate another proton. The ammonium ion (option E) cannot accept a further proton.

55. **(A)** The pressure at point A corresponds to the normal freezing and boiling points, so it must be 1.0 atm.

56. **(E)** The combustion of coal is simply the heat of formation of carbon dioxide.

57. **(D)** The solution with the lowest melting point will have the greatest number of particles dissolved in it. The 1.0 M magnesium chloride has 3 moles dissolved in it for each liter of solution. The next greatest number of dissolved particles would come from the sodium chloride and the hydrochloric acid, each of which would donate 2 moles of dissolved ions for each mole of compound.

58. **(C)** The heat absorbed by the water equals the mass of the water, multiplied by the specific heat of water and the increase in temperature change. The 840 J added to the water is enough heat to raise the temperature of 20 grams of water by 10 degrees.

59. **(A)** The iron reduction potential is the lowest, so it is most easily oxidized—not reduced. Since it is the most easily oxidized, then it acts as the reducing agent most easily.

60. **(C)** The quick way to look at this is that the carbon is 3/4 of the molecule by mass, so 12 grams (1 mole) of carbon must go with 4 grams (4 moles) of hydrogen. Or, to check with the complete calculation,

$$75 \text{ g C} \times \frac{1.0 \text{ mol C}}{12.011 \text{ g C}} = 6.2 \text{ mol C}$$

$$25 \text{ g H} \times \frac{1.0 \text{ mol H}}{1.008 \text{ g H}} = 24.8 \text{ mol H}$$

Empirical formula: CH_4

61. **(D)** The molar mass can be found by adding together the atomic weights of each atom in the formula: 14 g/mol (for nitrogen) + 3 × 1 g/mol (for the hydrogens) = 17 g/mol.

62. **(C)** To get a fractional van't Hoff factor, you should look for the molecule that is partially, but not completely ionized in water. The only molecule that fits that description is the weak acid, acetic acid.

63. **(D)**

$$3.5 \text{ g acid} = 1.0 \text{ L solution} \times \frac{0.10 \text{ mol acid}}{\text{L solution}} \times \frac{35 \text{ g acid}}{\text{mol acid}}$$

64. **(B)** 90 grams of water was removed upon heating, which is 3 times the 30 grams of crystalline solid that remained after heating.

65. **(D)** All these variables, except for the freezing point of the solute, are needed to calculate molar mass from freezing-point depression.

Molar mass can be calculated from the relationship for freezing-point depression:

$$\Delta T \text{ of the solution} = \text{molality of solute} \times k_f \text{ of the solution}$$

Once the molality is calculated, then the number of moles of unknown solute can be found because they equal the molality of the solution × kg solvent.

Finally, the molar mass of the unknown solute can be found by dividing the moles of the solute into the mass of the solute.

66. **(A)** Both the current and the length of time the current runs are needed to calculate the total amount of charge the electrodes receive. One mole of electrons is deposited for every 96,485 Coulombs of charge deposited at the electrode.

67. **(B)** At neutralization, the moles of acid equal the moles of base. There are $0.02 \times 0.1 = 0.002$ moles of base, so there must also be 0.002 moles of acid in the 0.5-liter sample; that would require an acid solution concentration of 0.004 moles per liter.

68. **(E)** The reaction rate is proportional to the square of the concentration of A, which means the reaction is second order with respect to A. The reaction rate does not change at all when the concentration of B changes, so it is zero order with respect to B. Combined, the reaction is second order overall.

69. **(E)**

$$240 \text{ C} = 1.0 \text{ A} \times 120 \text{ s} \times \frac{\text{C}}{\text{A s}}$$

70. **(C)** Acetylene shows a triple bond between the 2 carbons, which contains 2 pi bonds. In multiple bonds, the first bond is a sigma bond, where electrons are shared along the internuclear axis. Any additional bonding is created by the sideways overlap of unhybridized p-orbitals above and below the internuclear axis. None of the other options for answers contain multiple covalent bonds between any 2 atoms.

71. **(E)** None of these atoms change in oxidation state during the course of the reaction. Therefore, the change in oxidation state for aluminum is zero.

72. **(E)** Aluminum, with 3 electrons in its outer shell, will demonstrate the largest jump between its third and fourth ionization energies. Once it loses these 3 electrons, the fourth will need to come from electrons at $n = 2$.

73. **(C)** The compound with the greatest number of oxygen atoms bonded to the most electronegative central atom will pull most strongly on hydrogen's electron, thereby allowing the proton to be donated more easily.

74. **(A)** Only the size of the atom increases as you move down and to the left in the periodic table.

75. **(A)** Nickel has an unpaired electron in its d-orbital, which induces a magnetic field and causes it to be attracted to another magnetic field. Zinc chloride does not have such an unpaired electron and is not attracted to a magnet.

PRACTICE TEST 2

CLEP Chemistry

Also available at the REA Study Center (*www.rea.com/studycenter*)

This practice test is also offered online at the REA Study Center. Since all CLEP exams are computer-based, we recommend that you take the online version of the test to simulate test-day conditions and to receive these added benefits:

- **Timed testing conditions** – helps you gauge how much time you can spend on each question
- **Automatic scoring** – find out how you did on the test, instantly
- **On-screen detailed explanations of answers** – gives you the correct answer and explains why the other answer choices are wrong
- **Diagnostic score reports** – pinpoint where you're strongest and where you need to focus your study

PRACTICE TEST 2

CLEP Chemistry

(Answer sheets appear in the back of the book.)

TIME: 90 Minutes
75 Questions

PART A

Questions 1–4

(A) $Ag^+ + I^- \rightarrow AgI$
(B) $HC_2H_3O_2 + OH^- \rightarrow C_2H_3O_2{}^- + H_2O$
(C) $NaCl\ (l) \rightarrow Na\ (l) + Cl_2\ (g)$
(D) $Mg + O_2 \rightarrow MgO$
(E) $Fe_2{}^+ + 8H^+ + MnO_4{}^- \rightarrow Fe^{3+} + Mn^{2+} + 4H_2O$

1. Two electrons are transferred in this reaction.

2. This reaction results in a basic solution.

3. This reaction yields a cloudy product.

4. This reaction takes place in an electrolytic cell with the input of electrical energy.

Questions 5−9

    (A) H
    (B) F
    (C) Na
    (D) Th
    (E) Ar

5. The most electronegative element

6. The element that combines with hydrogen to form a halide

7. The most abundant element in the universe

8. The element whose most abundant naturally occurring isotopes are radioactive

9. The element that has a full outer shell

Questions 10−13

    (A) $N_2 (g) + 3H_2 (g) \rightarrow 2NH_3 (g)$
    (B) $AgI \rightarrow Ag^+ + I^-, \Delta H = -15$ kJ/mol
    (C) $H_2C_2O_4 + MnO_4^- + H2SO_4 \rightarrow MnSO_4 + CO_2 + H_2O$
    (D) $NaSCN \rightarrow Na^+ + SCN^-, \Delta H = 345$ kJ/mol
    (E) $C_4H_{10} + O_2 \rightarrow CO_2 + H_2O$

10. Increasing the temperature will increase the solubility of the ion in solution for this reaction.

11. Decreasing the temperature will increase the solubility of the ion in solution for this reaction.

12. This reaction is considered a combustion reaction.

13. Adding acid to this reaction will cause this reaction to go forward.

Questions 14−17

    (A) alpha decay
    (B) beta (β-) decay
    (C) positron decay
    (D) neutron capture
    (E) gamma radiation

14. The nuclear reaction that turns a proton into a neutron

15. The emission of very high energy photons

16. The nuclear reaction that gives off the nucleus of a helium atom

17. The nuclear reaction that turns a neutron into a proton

Questions 18−21

    (A) KI
    (B) H_2O_2
    (C) NaOH
    (D) $Fe(NO_3)_3$
    (E) Mg

18. Reaction of this substance with a strong acid would produce hydrogen gas

19. Heating this substance produces a gas

20. Combining a solution of this substance with a solution of lead nitrate will produce a yellow precipitate

21. Mixing this with KSCN will produce a blood-red solution

Use the following sets of thermodynamic parameters to match with questions 22–25.

  (A) H_2O
  (B) $C_{12}H_{26}$
  (C) PbS
  (D) Pb
  (E) SiO

22. To melt this substance, polar intermolecular attractions must be overcome.

23. To melt this substance, metallic attractions must be overcome.

24. To melt this substance, network covalent attractions must be overcome.

25. To melt this substance, ionic attractions must be overcome.

PART B

DIRECTIONS: Each of the questions or incomplete statements below is followed by five possible answers or completions. Select the best choice in each case and fill in the corresponding oval on the answer sheet.

26. The K_a for acetic acid is 1.8×10^{-5}. Calculate the approximate equilibrium constant for the reaction below.

$$HC_2H_3O_2 + OH^- \leftrightarrow H_2O + C_2H_3O_2^-$$

(A) 1.8×10^{-5}
(B) 1.8×10^9
(C) 1.0×10^{14}
(D) 5.5×10^4
(E) 1.0×10^{-14}

27. Use this table to answer the following question.

Substance	$S°$ (J/ K mol)
C_2H_2 (g)	249.0
CO_2 (g)	213.6
H_2O (l)	69.96
O_2 (g)	205.0

The numeric value for the standard change in entropy for the combustion of C_2H_2 (g) is closest to

(A) -267 J/mol K
(B) $+267$ J/mol K
(C) -497 J/mol K
(D) $+497$ J/mol K
(E) -763 J/mol K

28. Which of the following gases would be expected to show the greatest deviation from ideal behavior?

    (A) H_2
    (B) H_2S
    (C) H_2O
    (D) O_2
    (E) CH_4

29. The K_a of HCN is 1.0×10^{-10}. What is the approximate pH of a 1.0-molar solution of HCN?

    (A) 1
    (B) 3
    (C) 5
    (D) 7
    (E) 10

30. Use the equilibrium constants from the component reactions to calculate the equilibrium constant for the total reaction.

Reaction	Equilibrium Constant
$Cl_2 \rightarrow 2\ Cl$	K_1
$Cl + CO \rightarrow COCl$	K_2
$COCl + Cl \rightarrow COCl_2$	K_3
$Cl_2 + CO \rightarrow COCl_2$	K_{total}

    (A) $K_1 + K_2 + K_3$
    (B) $K_1 \times K_2 \times K_3$
    (C) $K_1 - K_2 - K_3$
    (D) $K_1 + K_2 - K_3$
    (E) $K_1 \times K_2 / K_3$

31. $Br_2 (l) \rightarrow Br_2 (g)$

The equilibrium vapor pressure of bromine gas is 0.281 atm at 25°C. What is the K_p for the above reaction at 25°C?
(A) 0 atm
(B) 0.281 atm
(C) $(0.281)^2$ atm
(D) 1.0 atm
(E) Cannot be determined from this data

32. Use the chart below to answer the following question.

Substance	ΔG_f° (kJ/mol)
$C_2 H_2 (g)$	209.0
$CO_2 (g)$	−394.0
$H_2O (l)$	−237.0
$O_2 (g)$	0.0

Which of the following represents the approximate change in standard free energy per mole for the combustion of $C_2H_2 (g)$?
(A) −20 kJ/mol
(B) +20 kJ/mol
(C) −100 kJ/mol
(D) −1200 kJ/mol
(E) +1200 kJ/mol

33. Which of the following sets of quantum numbers (listed in order of n, l, m_l, m_s) describes the highest energy valence electron of sodium in its ground state?

(A) 3, 0, 0, +½
(B) 3, 0, 0, −½
(C) 3, 0, 1, +½
(D) 3, 0, −1, −½
(E) 3, 1, 1, +½

34. When determining the molar volume of a gas by collecting the gas over water, which of the following laws must be considered to take into account the pressure exerted by water vapor in the collection bottle?

    (A) Boyle's law
    (B) Charles's law
    (C) Dalton's law
    (D) Henry's law
    (E) Pascal's law

35. Two carbon rods are put into a solution of $CuSO_4$. The electrodes are connected to a power source, which is turned on for 10 minutes. Which of the following will occur?

    (A) The mass of the cathode will increase.
    (B) The mass of the anode will increase.
    (C) Neither electrodes will change in mass.
    (D) A gas will be produced at the cathode.
    (E) No gas will be produced at either electrode.

36. Which of the following demonstrates the greatest increase between the atom's second and third ionization energies?

    (A) He
    (B) Na
    (C) Mg
    (D) Al
    (E) N

37. As you move across a row in the periodic table, which of the following traits does not increase?

    (A) Atomic radius
    (B) Electronegativity
    (C) Electron affinity
    (D) Ionization energy
    (E) Acidity of the oxides

38. Which of the following best explains why solid copper is a strong conductor, but solid copper sulfate is not?

    (A) The sulfate blocks the flow of electrons.
    (B) The unpaired electrons in copper's outer shell allow electricity to flow through the atom.
    (C) The *d*-level electrons in metallic bonds are delocalized.
    (D) Any compound with sulfur would not conduct electricity.
    (E) The copper is more diluted in copper sulfate.

39. Compare the bond angles in CH_4 versus XeF_4.

    (A) Both have the same bond angles.
    (B) CH_4 demonstrates 109.5° ; XeF_4 demonstrates 90°.
    (C) CH_4 demonstrates 109.5° ; XeF_4 demonstrates 120°.
    (D) CH_4 demonstrates 90° ; XeF_4 demonstrates 120°.
    (E) CH_4 demonstrates 90° ; XeF_4 demonstrates 109.5°.

40. Which of the following demonstrates the highest melting point?

    (A) Hydrogen gas
    (B) Butane
    (C) KI
    (D) SiC
    (E) The answer can't be determined from this information

41. Which of the following has the highest bond order?

    (A) Carbon tetrachloride
    (B) Carbon dioxide
    (C) Carbonate anion
    (D) Hydrogen gas
    (E) Water

42. The kinetics of a reaction in which two reactants, A and B, react together to form a product, C, is studied, and the results are recorded in the chart below. What is the order of the reaction with respect to A and B, respectively?

Trial	[A]	[B]	Initial rate of formation of C (M/s)
1	0.10	0.30	0.001
2	0.10	0.60	0.002
3	0.20	0.60	0.004

(A) First order with respect to A; first order with respect to B
(B) First order with respect to A; zero order with respect to B
(C) Zero order with respect to A; first order with respect to B
(D) Zero order with respect to A; second order with respect to B
(E) Second order with respect to A; zero order with respect to B

43. A sample of butane was completely combusted at STP. How many moles of butane were burned if 44.8 liters of carbon dioxide were produced as a result?

(A) 0.5 moles
(B) 1.0 moles
(C) 1.5 moles
(D) 2.0 moles
(E) 2.5 moles

44. Which of the following best describes the following reaction?

$$HC_2H_3O_2 + OH^- \rightarrow C_2H_3O_2^- + H_2O$$

(A) Oxidation-reduction reaction
(B) Reaction between a weak acid and a weak base
(C) Reaction between a strong acid and a strong base
(D) Reaction between a weak acid and a strong base
(E) Solvation reaction

45. If 2.0 moles of calcium carbonate is heated and decomposes in the following reaction, how many liters of carbon dioxide will be produced at STP?

$$CaCO_3 \ (s) \rightarrow CaO \ (g) + CO_2 \ (g)$$

(A) 12.2 L
(B) 22.4 L
(C) 44.8 L
(D) 67.2 L
(E) 89.6 L

46. Which of the following, when put into 1.0 M HCl, will NOT produce hydrogen gas?

(A) K
(B) Ca
(C) Mg
(D) Zn
(E) Cu

47. A mixture of hydrogen and carbon dioxide gas molecules is held in a container. How much faster will the hydrogen molecules be moving than the carbon dioxide molecules at any given temperature?

(A) They will move at the same speed.
(B) Hydrogen will move about twice as fast.
(C) Hydrogen will move about 3 times as fast.
(D) Hydrogen will move about 4 times as fast.
(E) Hydrogen will move about 5 times as fast.

48. Under what conditions does the nature of a gas begin to deviate from the ideal gas relationship?

(A) High temperatures and low pressures
(B) Low temperatures and high pressures
(C) High temperatures and high pressures
(D) Low temperatures and low pressures
(E) High kinetic energies and low densities

49. A weather balloon has a volume of 10.0 L at 1.0 atm on the ground. What will be its volume at a pressure of 0.25 atm, assuming constant temperature?

    (A) 1.25 L
    (B) 2.5 L
    (C) 5.0 L
    (D) 20.0 L
    (E) 40.0 L

50. A gas sample with a density of 1.6 g/L exerts a pressure of 1.0 atm at a temperature of 300 K. Which of the following best describes the molar mass of the gas?

    (A) $1.6 \times 0.08 \times 300$
    (B) 1.6×300
    (C) $1.6/300$
    (D) $300/1.6$
    (E) More information is needed to find the molar mass.

51. What do points A and B represent on the following phase diagram?

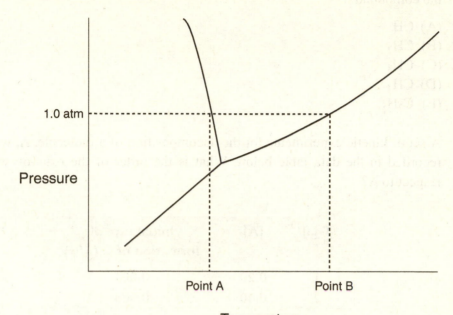

Temperature

(A) Point A is the normal freezing point; point B is the normal boiling point.
(B) Point A is the normal boiling point; point B is the normal freezing point.
(C) Point A is the normal boiling point; point B is the critical point.
(D) Point A is the triple point; point B is the critical point.
(E) Point A is the critical point; point B is the triple point.

52. What volume of nitrogen gas would be needed to react completely with 90 liters of hydrogen gas at STP in the following reaction?

$$N_2 (g) + 3 H_2 (g) \rightarrow 2 NH_3 (g)$$

(A) 10 L
(B) 20 L
(C) 30 L
(D) 60 L
(E) 120 L

53. A hydrocarbon contains 25% hydrogen. What is the empirical formula of the compound?

    (A) CH
    (B) CH_2
    (C) CH_3
    (D) CH_4
    (E) C_2H_5

54. A set of kinetic experiments for the decomposition of a molecule, A, were recorded in the data table below. What is the order of the reaction with respect to A?

Trial	[A]	Initial rate of formation of C (M/s)
1	0.20	0.001
2	0.40	0.004

    (A) Zero order
    (B) First order
    (C) Second order
    (D) Third order
    (E) Cannot be determined by the information given

55. Which of the following is NOT an assumption made by the kinetic molecular theory?

    (A) Gas particles collide with perfectly elastic collisions.
    (B) Gas particles are in constant, random straight-line motion.
    (C) Gas molecules exert no attractive force on each other.
    (D) The kinetic energy of the gas is directly proportional to its temperature.
    (E) The pressure exerted by the gas results from the gas colliding with the walls of the container.

56. Which of the following carries the units, J/g °C?

    (A) Enthalpy
    (B) Entropy
    (C) Free energy
    (D) Heat capacity
    (E) Heat of fusion

57. Which of the following orbitals, when filled, signify an especially stable atom?

    (A) *s* orbitals
    (B) *p* orbitals
    (C) *d* orbitals
    (D) *f* orbitals
    (E) Both *d* and *f* orbitals

58. The molecule or atom that can act either as an acid or a base is referred to as

    (A) coordinated
    (B) monoprotic
    (C) dibasic
    (D) amphoteric
    (E) ambiprotic

59. The equivalence point of an acid is reached in a titration when

    (A) the number of moles of acid molecules equals the number of moles of base molecules
    (B) the mass of the base added equals the mass of the acid that is present
    (C) the number of moles of added hydroxide ions equals the number of moles of protons
    (D) the amount of base that is added allows the acid to reach a neutral pH
    (E) the number of electrons lost equals the number of electrons gained

60. Which of the following adjusts to speed up a chemical reaction when a catalyst has been added?

    (A) The activation energy decreases.
    (B) The activation energy increases.
    (C) The reactant concentration increases.
    (D) The regular mechanism of the reaction speeds up.
    (E) The product concentration increases.

61. Oxidation of an atom in a reaction causes that atom to

    (A) lose control of some electrons
    (B) gain control of some electrons
    (C) have fewer oxygen atoms around it
    (D) be neutralized
    (E) precipitate

62. *Isomers* are defined as

    (A) two atoms that have the same number of protons, but a different number of neutrons
    (B) two atoms that have the same number of neutrons, but a different number of protons
    (C) two molecules that have the same molar mass, but have different physical properties
    (D) two atoms that have the same electrical conductivity
    (E) two molecules that have the same formula, but whose atoms are arranged differently

63. Which of the following laboratory procedures is used to separate two substances based on differences in their ability to dissolve in different solvents?

    (A) Titration
    (B) Centrifuge
    (C) Spectrophotometer
    (D) Distillation
    (E) Chromotography

64. Which of the following would be classified as an aldehyde?

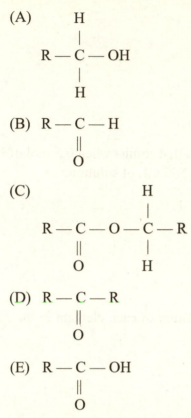

(A)

$$R - \overset{\overset{\displaystyle H}{|}}{\underset{\underset{\displaystyle H}{|}}{C}} - OH$$

(B) $R - \overset{C}{\underset{\underset{\displaystyle O}{\|}}{}} - H$

(C)

$$R - \overset{C}{\underset{\underset{\displaystyle O}{\|}}{}} - O - \overset{\overset{\displaystyle H}{|}}{\underset{\underset{\displaystyle H}{|}}{C}} - R$$

(D) $R - \overset{C}{\underset{\underset{\displaystyle O}{\|}}{}} - R$

(E) $R - \overset{C}{\underset{\underset{\displaystyle O}{\|}}{}} - OH$

65. A voltaic cell will demonstrate a standard total cell potential of 1.2 volts. Which of the following changes will result in a drop in the cell potential?

(A) An increase in the concentration of a soluble reactant
(B) An increase in the mass of a solid metal in the cell
(C) An increase in the concentration of a soluble product
(D) Turning the electricity off
(E) The dilution of all solutions in the cell

66. When the total cell voltage of a voltaic cell is greater than zero, which of the following is also true?

(A) K is less than 1 and ΔG is less than zero.
(B) K is greater than 1 and ΔG is less than zero.
(C) K is greater than 1 and ΔG is more than zero.
(D) K is less than 1 and ΔG is more than zero.
(E) K is zero and ΔG is greater than zero.

67. Which of the following orbitals take the highest-energy electrons of most transition metals?

    (A) s orbitals
    (B) p orbitals
    (C) d orbitals
    (D) f orbitals
    (E) both s and p orbitals

68. What is the approximate pH of a solution that results when 0.5 mol of HCl is completely dissolved in water to make 500 mL of solution?

    (A) 0
    (B) 0.5
    (C) 1.0
    (D) 2.0
    (E) 10.0

69. What is the approximate percent composition of each element in the compound potassium iodide?

    (A) 5% K, 95% I
    (B) 24% K, 76% I
    (C) 40% K, 60% I
    (D) 50% K, 50% I
    (E) 76% K, 24% I

70. What are the units of the specific rate constant for the reaction $A + B \rightarrow C$, which is represented by the kinetic data in the chart below?

Trial	[A]	[B]	Initial rate of formation of C (M/s)
1	0.20	0.30	0.02
2	0.20	0.60	0.04
3	0.40	0.60	0.08

    (A) $mol^2/L^2\ s$
    (B) $L^2/mol^2\ s$
    (C) $mol/L\ s$
    (D) $mol^2/L\ s$
    (E) $L/mol\ s$

71. Which of the following atoms are available to form expanded octets?

    (A) Na
    (B) N
    (C) O
    (D) C
    (E) P

72. The difference between using freezing-point depression to find the molar mass of an electrolyte and using the same method to find the molar mass of a nonelectrolyte is that when dealing with an electrolyte, the student also must know

    (A) the van't Hoff factor for the known solvent
    (B) the van't Hoff factor for the unknown electrolyte
    (C) the van't Hoff factor for the nonelectrolyte
    (D) the molarity of the pure electrolyte
    (E) There is no difference between an electrolyte and a nonelectrolyte in this calculation

73. A student cut a 5.30 g magnesium ribbon into small pieces and then began to heat it in a 16.30 g crucible. Over time, the magnesium ignited and burned to form a white, grayish and powdery substance. After the crucible cooled, the student measured the mass of the crucible that contained the powder and found it to be 24.80 g. What was the mass of the resulting oxide?

    (A) 5.30 g
    (B) 8.50 g
    (C) 12.50 g
    (D) 24.80 g
    (E) cannot be determined from this information

74. A student heated a 90-gram sample of a hydrated salt. When it was heated to the point when all the water was removed, the remaining powder weighed 60 grams. What is the ratio of the mass of salt to the mass of water in the hydrate?

    (A) 1:1
    (B) 3:2
    (C) 2:1
    (D) 2:3
    (E) 1:2

75. In the following unbalanced reaction, once the reaction is balanced, how many moles of hydrogen ions are needed to react with each mole of aluminum hydroxide?

$$Al(OH)_3 + H^+ \rightarrow Al^{3+} + H_2O$$

(A) 1
(B) 2
(C) 3
(D) 4
(E) 5

PRACTICE TEST 2

Answer Key

1.	(D)	26.	(B)	51.	(A)
2.	(B)	27.	(A)	52.	(C)
3.	(A)	28.	(B)	53.	(D)
4.	(C)	29.	(C)	54.	(C)
5.	(B)	30.	(B)	55.	(C)
6.	(C)	31.	(B)	56.	(D)
7.	(A)	32.	(D)	57.	(B)
8.	(D)	33.	(A)	58.	(D)
9.	(E)	34.	(C)	59.	(C)
10.	(D)	35.	(A)	60.	(A)
11.	(B)	36.	(C)	61.	(A)
12.	(E)	37.	(A)	62.	(E)
13.	(C)	38.	(C)	63.	(E)
14.	(C)	39.	(B)	64.	(B)
15.	(E)	40.	(D)	65.	(C)
16.	(A)	41.	(B)	66.	(B)
17.	(B)	42.	(A)	67.	(C)
18.	(E)	43.	(A)	68.	(A)
19.	(B)	44.	(D)	69.	(B)
20.	(A)	45.	(C)	70.	(E)
21.	(D)	46.	(E)	71.	(E)
22.	(A)	47.	(E)	72.	(B)
23.	(D)	48.	(B)	73.	(B)
24.	(E)	49.	(E)	74.	(C)
25.	(C)	50.	(A)	75.	(C)

PRACTICE TEST 2

Detailed Explanations of Answers

1. **(D)** Two electrons are transferred per atom in the oxidation-reduction reaction between magnesium and oxygen. Magnesium is oxidized by losing two electrons, and oxygen is reduced.

2. **(B)** The titration of acetic acid by a strong base results in the production of water and the acetate ion, which is a weak base—thereby leaving the solution basic.

3. **(A)** A cloudy product results in the formation of silver iodide (AgI) precipitate.

4. **(C)** In an electrolytic cell, melted sodium chloride will react by having the sodium ions move toward the cathode and the chloride ion move toward the anode. Sodium metal will be formed at the cathode, and chloride gas will be formed at the anode.

5. **(B)** Flourine is the most electronegative element. Its tendency to gain an electron to have a full set of p-orbitals, combined with the small atomic radius which puts the outer electrons close to the positively charged nucleus, make fluorine highly electronegative.

6. **(C)** Sodium, along with other alkaline metals, will combine with hydrogen to form halide compounds. Halides are unique in that they are the only time that hydrogen atoms take a negative oxidation state.

7. **(A)** Hydrogen is the most abundant element in the universe; oxygen is the most abundant element on the Earth.

8. **(D)** Thorium (Th) is the heaviest element, and the only one whose most abundant naturally occurring isotopes are radioactive. While the others have naturally occurring isotopes that are radioactive, such isotopes occur with very low frequency.

9. **(E)** Argon, as an inert gas, has a full set of p-orbitals and s-orbitals at $n = 3$.

10. **(D)** Solubility increases when the temperature is raised on endothermic solution reactions. According to Le Chatelier's principle, the reaction will respond by using the heat and dissolving more into solution.

11. **(B)** Solubility decreases when the temperature is raised on an exothermic solution reaction.

12. **(E)** The oxidation of butane to form carbon dioxide and water is considered a combustion reaction.

13. **(C)** Adding an acid to this reaction will prompt a forward reaction as a result of Le Chatelier's principle. The reaction will respond by using up the acid, which is chemically the same as adding sulfuric acid.

14. **(C)** Positron decay results in the departure of a subatomic particle that carries away a positive charge and results in a proton being turned into a neutron. As a result, the mass number remains the same, but the atomic number decreases by 1.

15. **(E)** Gamma radiation is composed of photons of very short wavelength and high energy. Gamma radiation usually accompanies other types of radioactive decay and is the vehicle by which energy is carried away from the atom.

16. **(A)** Alpha decay results in a bundle of nuclear material leaving the atom. This is a very stable bundle of material that consists of 2 protons and 2 neutrons—equivalent to the nucleus of a helium atom.

17. **(B)** Beta decay is the opposite of positron decay and results in an electron carrying a negative charge away from the atom and ultimately turning a neutron into a proton. The overall mass number stays the same, but the atomic number increases by 1.

18. **(E)** Magnesium is an active metal that will release its electrons to protons to produce hydrogen gas. The other metals that are options are bound up in ionic attractions and have already released control of their electrons.

19. **(B)** Heating hydrogen peroxide will increase the rate at which it decomposes into oxygen gas and water.

20. **(A)** Potassium iodide will dissolve and yield soluble iodide ions, which will combine with lead ions in solution to form the yellow precipitate lead iodide.

21. **(D)** The combination of the thiocyanite ion with the soluble iron (III) ion yields blood-red iron (III) thiocyanite, a coordination compound.

22. **(A)** The crystal structure of frozen water, or ice, is held together from the polar intermolecular attractions between the slightly positive hydrogen side of the molecule *and* the slightly negative oxygen side of other water molecules.

23. **(D)** The crystal structure of solid metal is held together by metallic attractions, which must be overcome if the metal is to melt.

24. **(E)** The SiO crystal is held together with network covalent forces, which are the strongest of interatomic forces.

25. **(C)** PbS is held together with ionic attractions, which will need to be overcome in order to melt that substance.

26. **(B)** The reaction as shown is the reverse of the K_b reaction. The equilibrium constant for this reaction is simply the K_a divided by K_w, or simply add 14 to the exponent of the acid dissociation constant.

27. **(A)**

$$C_2H_2 + \frac{5}{2}O_2 \rightarrow 2CO_2 + H_2O$$

$\Delta S°_{reaction} = \Sigma S°\ (\text{products}) - \Sigma S°\ (\text{reactants})$

$\Delta S°_{reaction} = [(2 \times 213.6) + (70)] - [(249) + (2.5 \times 205)]$

$\Delta S°_{reaction} = -264.3$ J/K

28. **(B)** The ideal gas law assumes that there is no intermolecular attraction and that the volume of the gas molecules is insignificant relative to the space between the molecules. Therefore, gases that are large molecules and/or demonstrate strong intermolecular attraction are likely to be nonideal gases. Hydrogen sulfide is both the largest gas molecule of the options and shows very strong intermolecular attraction because of the difference in electronegativity between hydrogen and sulfur. Water vapor also shows strong intermolecular attraction, but it is a much smaller molecule.

29. **(C)** The square root of the ionization constant will give the molar proton concentration (10^{-5}). The negative of the exponent of the proton concentration gives the pH, or 5.

30. **(B)** When reactions are added together to get a total reaction, the equilibrium constants of the component reactions are multiplied together to get the equilibrium constant for the total reaction.

31. **(B)** The gaseous equilibrium constant is composed of only those compounds that are in the gas phase. For a process that only involves a single gas in the reaction—a phase change from liquid to gas, for example—the

equilibrium expression consists only of the pressure of the gas at equilibrium, which is the definition of the equilibrium vapor pressure of the gas.

32. **(D)**

$$C_2H_2 + \frac{5}{2}O_2 \rightarrow 2CO_2 + H_2O$$

$\Delta G°_{reaction} = \Sigma \Delta G_f \text{ (products)} - \Sigma \Delta G°_f \text{ (reactants)}$

$\Delta G°_{reaction} = [(2 \times -394) + (-237)] - (209)$

$\Delta G°_{reaction} = -1,234 \text{ kJ}$

33. **(A)** The highest energy electron in sodium is in the third energy level, or $n = 3$—the first quantum number. The second 2 quantum numbers are 0 to depict the s-orbital. The first electron in that orbital is denoted with $+1/2$ spin. This gives that highest energy electron the quantum numbers of 3, 0, 0, $+1/2$.

34. **(C)** If the collection vessel is exposed to the surface of water, as it usually is when collecting a gas in the laboratory, then water evaporates from the surface and combines with the collected gas. Therefore, to be accurate in calculating how many moles of gas are produced in the reaction, Dalton's law must be used to account for the moles of water vapor also in the collection vessel. Dalton's law states that the total pressure exerted by a mixture of gases equals the sum of the partial pressures of each component in the mixture.

35. **(A)** The mass of the cathode will increase because the copper ions in solution will gain electrons and precipitate solid copper onto the cathode.

36. **(C)** Magnesium demonstrates the greatest difference between its second and third ionization energies because it has 2 electrons in its outer shell. To remove the third would require breaking into the inner, more tightly held, second shell—which requires considerably more energy.

37. **(A)** Moving to the right on the periodic table across a row, each new element experiences a greater pull between the nucleus and the ever-increasing number of electrons, which pulls in the outer shell of electrons and decreases the atomic radius.

38. **(C)** The delocalized electrons in copper and other transition metals create an "electron sea," which is fundamental to the construction of the metallic attraction that holds solid copper together.

39. **(B)** CH_4 demonstrates 109.5° between the C-H bonds, while XeF_4 demonstrates 90° angles between its Xe-F bonds. Even though the formulas are similar, the different geometry is a result of the larger size of the Xe atom and the highly electronegative F atoms creating an "expanded octet." Two other pairs of unbonded electron pairs around Xenon also take up space, forcing the bonds to fluorine to be closer together than the C-H bonds in methane.

40. **(D)** The highest melting point will be demonstrated by the crystal that has the strongest forces holding it in the crystalline structure. In this case, SiC is held together with network covalent attractions, which are stronger than ionic and polar intermolecular attractions.

41. **(B)** Carbon dioxide has a bond order of 2.0 (double bonds) between the carbon and the oxygen. Other than the carbonate ion, the other molecules have all single bonds. The carbonate ion demonstrates resonance, with a delocalized pair of electrons shared across 3 carbon-oxygen bonds, which results in a bond order of about 1.3.

42. **(A)** The reaction is first order with respect to both A and B. This can be seen in comparing two experiments where the concentration of one reactant is constant and the other reactant is doubled. If the reaction rate doubles, then the rate increases in direct proportion to the concentration and shows first-order kinetics.

43. **(A)** In the balanced reaction for the combustion of butane, 4 moles of carbon dioxide are produced for every mole of butane used up. 44.8 liters of gas represents 20 moles at STP, which therefore uses 0.5 mole of butane.

44. **(D)** This reaction is the neutralization of a weak acid, acetic acid, by a strong base. This is a Brønsted-Lowry acid-base reaction, which yields water and the soluble acetate ion.

45. **(C)** For every 2 moles of calcium carbonate that decomposes, 2 moles of carbon dioxide will be produced. Two moles of gas at STP will occupy $2 \times 22.4 \text{ L} = 44.8 \text{ L}$.

46. **(E)** Only copper is less reactive than acidic protons, but all the others will yield electrons to protons to form hydrogen gas.

47. **(E)** The molar mass of carbon dioxide is 24 times the molar mass of hydrogen gas. The velocity of the molecule is proportional to the inverse of the square of the molar mass. Therefore, hydrogen gas molecules move about 5 times as fast as carbon dioxide.

48. **(B)** Deviations occur in ideal gas behavior when gas particles get very close together, which is at low temperatures and high pressures. Under these conditions, intermolecular attraction and molecular volume can become significant, whereas these variables are otherwise neglected in the ideal gas law.

49. **(E)** Since pressure is inversely proportional to volume, decreasing pressure to a quarter of its previous level will result in a 4-fold increase in volume—assuming constant temperature.

50. **(A)** The molar mass of a gas may be calculated from the version of the ideal gas law where molar mass = density $\times$ R $\times$ temperature/pressure.

51. **(A)** The normal freezing and boiling points on a phase diagram are represented by the point on the vapor-pressure curve and liquid-solid curve that correspond to 1.0 atm pressure.

52. **(C)** By the law of combining volumes, the volumes of gases in a reaction are proportional to the molar volumes. The molar ratio of 3:1 between hydrogen and nitrogen gases would indicate that the same ratio would be true for gas volume. It would take 30 L of nitrogen gas to react with 90 L of hydrogen gas.

53. **(D)** If a hydrocarbon is 25% hydrogen by mass, then it is 75% carbon, which corresponds to 1 mole of carbon (at approximately 12 g/mol) combining with 4 moles of hydrogen (at approximately 1 g/mol each).

    Another way to consider this problem is to assume that there is 100 g of material, and 25 grams of that is hydrogen and 75 grams is carbon. Converting each of these values to molar amounts will result in a 4:1 hydrogen/carbon ratio.

54. **(C)** The velocity of the reaction quadruples when the concentration of A doubles. Therefore, the exponent of [A] in the rate equation is 2, and it is second order with respect to A.

55. **(C)** The assumption that gas molecules do not exert any attractive force on each other is an assumption of the ideal gas law, not the kinetic molecular theory.

56. **(D)** Heat capacity is the amount of heat that will raise 1 gram of substance by 1 degree. Option A (entropy) is a measure of disorder, which is an amount of heat that is lost as waste heat per degree of temperature (J/K). Options B, C, and E are all types of energy, which is measured in joules (J).

57. **(B)** While any set of filled orbitals gives an atom some level of stability, a filled set of *p*-orbitals confers a higher level of chemical stability.

58. **(D)** *Amphoterism* is the name given to atoms or molecules that can either accept or donate protons, or act as either an acid or a base.

59. **(C)** The equivalence point does not depend on reaching a neutral pH (as is the case with a strong base/weak acid titration that reaches equivalence at a basic pH). It does not depend on having the same number of moles of acid molecules as base molecules (proved by titrating sodium hydroxide with sulfuric acid). The equivalence point is reached in a titration when the number of moles of hydroxide ions in the solution matches the number of moles of protons in the solution.

60. **(A)** The addition of a catalyst will cause the reaction to undergo a faster mechanism that has a different, lower activation energy.

61. **(A)** Oxidation of an atom causes its oxidation number to increase. The oxidation number for an atom is higher when it controls fewer electrons.

62. **(E)** Isomers are 2 or more molecules that have the same formula, but the atoms in the 2 molecules have a different arrangement in space. This often results in a difference in the physical properties of the 2 molecules, but may not necessarily be the case.

63. **(E)** Chromatography is a procedure that relies on the solubility attraction between components in a mixture with stationary and mobile phases. The components in the mixture that adhere better to the mobile phase will travel with it as it passes through the stationary phase. As a result of the competing attractions based on solubility, components in the mixture are separated.

64. **(B)** An aldehyde is a functional group in which a carbon at the end of a carbon chain, or not bonded to any other carbons, has a bond to a hydrogen and a double bond to an oxygen atom.

65. **(C)** The voltage in a voltaic cell will be equal to the standard voltage as long as all concentrations continue to be 1.0 molar and pressures equal to 1.0 atm pressure. If the concentrations of the soluble reactants drop below 1.0 molar, or the concentrations of the soluble products exceed 1.0 molar, then the voltage in the cell will drop.

66. **(B)** When a cell voltage is greater than 1.0, it is a spontaneous reaction that favors the formation of products. This would correspond with an equilibrium constant greater than 1.0 and a change in free energy less than 0. These relationships are summarized by the following equation:

$$\Delta G° = -nFE°_{total} = -RT \ln K$$

67. **(C)** All transition metals have their outer electrons in *s*-orbitals, but their highest energy electron is in an inner *d*-orbital.

68. **(A)** pH is the negative logarithm of the hydrogen ion concentration. For example, $[H^+] = 0.1$ corresponds to a pH of 1; a $[H^+] = 1.0$ (as in this question) corresponds to a pH of 0.

69. **(B)** Potassium iodide has the formula KI—a 1:1 molar ratio. However, the mass of iodine is approximately 3 times the mass of potassium, so the iodine has 3 times the contribution to the percent mass.

This problem may also be solved by dividing the molar mass of both potassium and iodine into the total molar mass and then multiplying by 100.

70. **(E)** Given the data in this problem, the rate law is first order with respect to both A and B, or second order overall. (This is true because each causes the reaction rate to double when the concentration is doubled.) To find the units of the specific rate constant (which are different for rate laws of different order), plug the known units into the rate equation.

$$\text{Reaction rate} = k\,[A]\,[B]$$

$$\text{Mol/L s} = k\,[\text{mol/L}]\,[\text{mol/L}]$$

$$k = \text{L/mol s}$$

71. **(E)** Expanded octets are potentially available to atoms that have electrons in *d*-orbitals. Phosphorus is the only atom with *d*-orbitals.

72. **(B)** The van't Hoff factor depicts the degree of dissociation of the solute. A 20% dissociation results in a van't Hoff factor of 1.2. This information is critical when finding molar mass using freezing-point depression, since the molarity of the solution will be inversely proportional to the van't Hoff factor.

73. **(B)** The mass of the remaining oxide equals the difference between the mass of the crucible and the mass and crucible of the oxide together after heating.

74. **(C)** The salt alone weighs 60 grams, and the water removed from the hydrated salt weighs 30 grams. Therefore, the mass of the salt is twice the mass of the water in the hydrate.

75. **(C)** The balanced reaction is as follows:

$$Al(OH)_3 + 3\,H^+ \rightarrow Al^{3+} + 3\,H_2O$$

ANSWER SHEETS

Practice Test 1
Practice Test 2

PRACTICE TEST 1

Answer Sheet

1. Ⓐ Ⓑ Ⓒ Ⓓ Ⓔ	26. Ⓐ Ⓑ Ⓒ Ⓓ Ⓔ	51. Ⓐ Ⓑ Ⓒ Ⓓ Ⓔ
2. Ⓐ Ⓑ Ⓒ Ⓓ Ⓔ	27. Ⓐ Ⓑ Ⓒ Ⓓ Ⓔ	52. Ⓐ Ⓑ Ⓒ Ⓓ Ⓔ
3. Ⓐ Ⓑ Ⓒ Ⓓ Ⓔ	28. Ⓐ Ⓑ Ⓒ Ⓓ Ⓔ	53. Ⓐ Ⓑ Ⓒ Ⓓ Ⓔ
4. Ⓐ Ⓑ Ⓒ Ⓓ Ⓔ	29. Ⓐ Ⓑ Ⓒ Ⓓ Ⓔ	54. Ⓐ Ⓑ Ⓒ Ⓓ Ⓔ
5. Ⓐ Ⓑ Ⓒ Ⓓ Ⓔ	30. Ⓐ Ⓑ Ⓒ Ⓓ Ⓔ	55. Ⓐ Ⓑ Ⓒ Ⓓ Ⓔ
6. Ⓐ Ⓑ Ⓒ Ⓓ Ⓔ	31. Ⓐ Ⓑ Ⓒ Ⓓ Ⓔ	56. Ⓐ Ⓑ Ⓒ Ⓓ Ⓔ
7. Ⓐ Ⓑ Ⓒ Ⓓ Ⓔ	32. Ⓐ Ⓑ Ⓒ Ⓓ Ⓔ	57. Ⓐ Ⓑ Ⓒ Ⓓ Ⓔ
8. Ⓐ Ⓑ Ⓒ Ⓓ Ⓔ	33. Ⓐ Ⓑ Ⓒ Ⓓ Ⓔ	58. Ⓐ Ⓑ Ⓒ Ⓓ Ⓔ
9. Ⓐ Ⓑ Ⓒ Ⓓ Ⓔ	34. Ⓐ Ⓑ Ⓒ Ⓓ Ⓔ	59. Ⓐ Ⓑ Ⓒ Ⓓ Ⓔ
10. Ⓐ Ⓑ Ⓒ Ⓓ Ⓔ	35. Ⓐ Ⓑ Ⓒ Ⓓ Ⓔ	60. Ⓐ Ⓑ Ⓒ Ⓓ Ⓔ
11. Ⓐ Ⓑ Ⓒ Ⓓ Ⓔ	36. Ⓐ Ⓑ Ⓒ Ⓓ Ⓔ	61. Ⓐ Ⓑ Ⓒ Ⓓ Ⓔ
12. Ⓐ Ⓑ Ⓒ Ⓓ Ⓔ	37. Ⓐ Ⓑ Ⓒ Ⓓ Ⓔ	62. Ⓐ Ⓑ Ⓒ Ⓓ Ⓔ
13. Ⓐ Ⓑ Ⓒ Ⓓ Ⓔ	38. Ⓐ Ⓑ Ⓒ Ⓓ Ⓔ	63. Ⓐ Ⓑ Ⓒ Ⓓ Ⓔ
14. Ⓐ Ⓑ Ⓒ Ⓓ Ⓔ	39. Ⓐ Ⓑ Ⓒ Ⓓ Ⓔ	64. Ⓐ Ⓑ Ⓒ Ⓓ Ⓔ
15. Ⓐ Ⓑ Ⓒ Ⓓ Ⓔ	40. Ⓐ Ⓑ Ⓒ Ⓓ Ⓔ	65. Ⓐ Ⓑ Ⓒ Ⓓ Ⓔ
16. Ⓐ Ⓑ Ⓒ Ⓓ Ⓔ	41. Ⓐ Ⓑ Ⓒ Ⓓ Ⓔ	66. Ⓐ Ⓑ Ⓒ Ⓓ Ⓔ
17. Ⓐ Ⓑ Ⓒ Ⓓ Ⓔ	42. Ⓐ Ⓑ Ⓒ Ⓓ Ⓔ	67. Ⓐ Ⓑ Ⓒ Ⓓ Ⓔ
18. Ⓐ Ⓑ Ⓒ Ⓓ Ⓔ	43. Ⓐ Ⓑ Ⓒ Ⓓ Ⓔ	68. Ⓐ Ⓑ Ⓒ Ⓓ Ⓔ
19. Ⓐ Ⓑ Ⓒ Ⓓ Ⓔ	44. Ⓐ Ⓑ Ⓒ Ⓓ Ⓔ	69. Ⓐ Ⓑ Ⓒ Ⓓ Ⓔ
20. Ⓐ Ⓑ Ⓒ Ⓓ Ⓔ	45. Ⓐ Ⓑ Ⓒ Ⓓ Ⓔ	70. Ⓐ Ⓑ Ⓒ Ⓓ Ⓔ
21. Ⓐ Ⓑ Ⓒ Ⓓ Ⓔ	46. Ⓐ Ⓑ Ⓒ Ⓓ Ⓔ	71. Ⓐ Ⓑ Ⓒ Ⓓ Ⓔ
22. Ⓐ Ⓑ Ⓒ Ⓓ Ⓔ	47. Ⓐ Ⓑ Ⓒ Ⓓ Ⓔ	72. Ⓐ Ⓑ Ⓒ Ⓓ Ⓔ
23. Ⓐ Ⓑ Ⓒ Ⓓ Ⓔ	48. Ⓐ Ⓑ Ⓒ Ⓓ Ⓔ	73. Ⓐ Ⓑ Ⓒ Ⓓ Ⓔ
24. Ⓐ Ⓑ Ⓒ Ⓓ Ⓔ	49. Ⓐ Ⓑ Ⓒ Ⓓ Ⓔ	74. Ⓐ Ⓑ Ⓒ Ⓓ Ⓔ
25. Ⓐ Ⓑ Ⓒ Ⓓ Ⓔ	50. Ⓐ Ⓑ Ⓒ Ⓓ Ⓔ	75. Ⓐ Ⓑ Ⓒ Ⓓ Ⓔ

PRACTICE TEST 2

Answer Sheet

1. Ⓐ Ⓑ Ⓒ Ⓓ Ⓔ	26. Ⓐ Ⓑ Ⓒ Ⓓ Ⓔ	51. Ⓐ Ⓑ Ⓒ Ⓓ Ⓔ
2. Ⓐ Ⓑ Ⓒ Ⓓ Ⓔ	27. Ⓐ Ⓑ Ⓒ Ⓓ Ⓔ	52. Ⓐ Ⓑ Ⓒ Ⓓ Ⓔ
3. Ⓐ Ⓑ Ⓒ Ⓓ Ⓔ	28. Ⓐ Ⓑ Ⓒ Ⓓ Ⓔ	53. Ⓐ Ⓑ Ⓒ Ⓓ Ⓔ
4. Ⓐ Ⓑ Ⓒ Ⓓ Ⓔ	29. Ⓐ Ⓑ Ⓒ Ⓓ Ⓔ	54. Ⓐ Ⓑ Ⓒ Ⓓ Ⓔ
5. Ⓐ Ⓑ Ⓒ Ⓓ Ⓔ	30. Ⓐ Ⓑ Ⓒ Ⓓ Ⓔ	55. Ⓐ Ⓑ Ⓒ Ⓓ Ⓔ
6. Ⓐ Ⓑ Ⓒ Ⓓ Ⓔ	31. Ⓐ Ⓑ Ⓒ Ⓓ Ⓔ	56. Ⓐ Ⓑ Ⓒ Ⓓ Ⓔ
7. Ⓐ Ⓑ Ⓒ Ⓓ Ⓔ	32. Ⓐ Ⓑ Ⓒ Ⓓ Ⓔ	57. Ⓐ Ⓑ Ⓒ Ⓓ Ⓔ
8. Ⓐ Ⓑ Ⓒ Ⓓ Ⓔ	33. Ⓐ Ⓑ Ⓒ Ⓓ Ⓔ	58. Ⓐ Ⓑ Ⓒ Ⓓ Ⓔ
9. Ⓐ Ⓑ Ⓒ Ⓓ Ⓔ	34. Ⓐ Ⓑ Ⓒ Ⓓ Ⓔ	59. Ⓐ Ⓑ Ⓒ Ⓓ Ⓔ
10. Ⓐ Ⓑ Ⓒ Ⓓ Ⓔ	35. Ⓐ Ⓑ Ⓒ Ⓓ Ⓔ	60. Ⓐ Ⓑ Ⓒ Ⓓ Ⓔ
11. Ⓐ Ⓑ Ⓒ Ⓓ Ⓔ	36. Ⓐ Ⓑ Ⓒ Ⓓ Ⓔ	61. Ⓐ Ⓑ Ⓒ Ⓓ Ⓔ
12. Ⓐ Ⓑ Ⓒ Ⓓ Ⓔ	37. Ⓐ Ⓑ Ⓒ Ⓓ Ⓔ	62. Ⓐ Ⓑ Ⓒ Ⓓ Ⓔ
13. Ⓐ Ⓑ Ⓒ Ⓓ Ⓔ	38. Ⓐ Ⓑ Ⓒ Ⓓ Ⓔ	63. Ⓐ Ⓑ Ⓒ Ⓓ Ⓔ
14. Ⓐ Ⓑ Ⓒ Ⓓ Ⓔ	39. Ⓐ Ⓑ Ⓒ Ⓓ Ⓔ	64. Ⓐ Ⓑ Ⓒ Ⓓ Ⓔ
15. Ⓐ Ⓑ Ⓒ Ⓓ Ⓔ	40. Ⓐ Ⓑ Ⓒ Ⓓ Ⓔ	65. Ⓐ Ⓑ Ⓒ Ⓓ Ⓔ
16. Ⓐ Ⓑ Ⓒ Ⓓ Ⓔ	41. Ⓐ Ⓑ Ⓒ Ⓓ Ⓔ	66. Ⓐ Ⓑ Ⓒ Ⓓ Ⓔ
17. Ⓐ Ⓑ Ⓒ Ⓓ Ⓔ	42. Ⓐ Ⓑ Ⓒ Ⓓ Ⓔ	67. Ⓐ Ⓑ Ⓒ Ⓓ Ⓔ
18. Ⓐ Ⓑ Ⓒ Ⓓ Ⓔ	43. Ⓐ Ⓑ Ⓒ Ⓓ Ⓔ	68. Ⓐ Ⓑ Ⓒ Ⓓ Ⓔ
19. Ⓐ Ⓑ Ⓒ Ⓓ Ⓔ	44. Ⓐ Ⓑ Ⓒ Ⓓ Ⓔ	69. Ⓐ Ⓑ Ⓒ Ⓓ Ⓔ
20. Ⓐ Ⓑ Ⓒ Ⓓ Ⓔ	45. Ⓐ Ⓑ Ⓒ Ⓓ Ⓔ	70. Ⓐ Ⓑ Ⓒ Ⓓ Ⓔ
21. Ⓐ Ⓑ Ⓒ Ⓓ Ⓔ	46. Ⓐ Ⓑ Ⓒ Ⓓ Ⓔ	71. Ⓐ Ⓑ Ⓒ Ⓓ Ⓔ
22. Ⓐ Ⓑ Ⓒ Ⓓ Ⓔ	47. Ⓐ Ⓑ Ⓒ Ⓓ Ⓔ	72. Ⓐ Ⓑ Ⓒ Ⓓ Ⓔ
23. Ⓐ Ⓑ Ⓒ Ⓓ Ⓔ	48. Ⓐ Ⓑ Ⓒ Ⓓ Ⓔ	73. Ⓐ Ⓑ Ⓒ Ⓓ Ⓔ
24. Ⓐ Ⓑ Ⓒ Ⓓ Ⓔ	49. Ⓐ Ⓑ Ⓒ Ⓓ Ⓔ	74. Ⓐ Ⓑ Ⓒ Ⓓ Ⓔ
25. Ⓐ Ⓑ Ⓒ Ⓓ Ⓔ	50. Ⓐ Ⓑ Ⓒ Ⓓ Ⓔ	75. Ⓐ Ⓑ Ⓒ Ⓓ Ⓔ

Glossary

accuracy: a measure of how closely individual measurements agree with the correct value.

acid: a substance that is able to donate an H^+ ion (proton) when dissolved in water.

activated complex: a short-lived, high-energy arrangement of atoms that is found in an energy diagram; often referred to as the transition state between reactants and products.

activation energy: the minimum amount of energy required in order for a reaction to proceed from reactants to products.

alcohol: an organic molecule containing an –OH as a functional group in place of a hydrogen.

aldehyde: an organic molecule containing the –CHO functional group.

alkali metals: elements that are located in Group IA of the periodic table.

alkaline earth metals: elements that are located in Group IIA of the periodic table.

alkanes: type of hydrocarbon having only single bonds; follows the formula C_nH_{2n+2}.

alkenes: type of hydrocarbon having a double bond; follows the formula C_nH_{2n}.

alkynes: type of hydrocarbon having a triple bond; follows the formula C_nH_{2n-2}.

alloy: a homogeneous mixture of metals.

alpha decay: a type of radioactive decay that occurs for radioactive elements having an atomic number greater than 82, resulting in the emission of an alpha particle.

alpha particle: a particle equivalent to a helium nucleus consisting of two protons and two neutrons, ^4_2He.

amine: an organic molecule containing the –NH_2 as a functional group.

amorphous solids: a solid whose molecular arrangement of atoms lack a pattern, resulting in an undefined shape.

amphoteric: a substance that can act as either an acid or a base.

angular momentum quantum number: the second of four quantum numbers, which indicates the type of orbital within an energy level.

anion: a negatively charged ion formed by the gain of electrons by an atom.

anode: electrode in an electrochemical cell where oxidation takes place.

aqueous: solution in which water is the solvent.

Arrhenius theory: theory of acid–base chemistry where an acid is defined as a substance that produces H_3O^+ and a base produces OH^-.

atom: smallest building block of matter that still retains its properties; composed of subatomic particles called protons, neutrons, and electrons.

atomic mass: the weighted average of the mass numbers of individual isotopes of an atom.

atomic number: the number of protons found in an element; distinguishes one element from another.

Aufbau principle: the principle that electrons will fill atomic orbitals from lowest energy to highest energy.

Avogadro's law: statement that indicates that gases of equal volume contain the same number of molecules at the same temperature and pressure.

base: substance that is an H^+ acceptor or produces excess OH^-.

beta decay: radioactive decay process in which a nucleus emits a high-energy electron resulting from a neutron being converted to a proton.

body-centered crystal: cubic cell in which the lattice points occur at the corners and center.

boiling point: temperature at which the vapor pressure of a liquid equals the atmospheric pressure.

bomb calorimeter: device for measuring the heat released in a combustion reaction under constant volume.

bond energy: amount of energy contained within a chemical bond.

Boyle's law: law stating that at constant temperature, volume is inversely proportional to pressure.

Brønsted-Lowry theory: states that acids will act as proton donors and bases will be proton acceptors.

buffer: a solution of a weak acid and its conjugate or a weak base and its conjugate whose pH will not change appreciably when additional acid or base is added to the solution.

calorie: unit of energy; amount of energy required to raise the temperature of 1 g of water by 1 degree Celsius.

calorimeter: device used to measure the energy change in a chemical reaction; based on the law of conservation of energy.

carbonyl group: an organic functional group where a carbon atom is bonded to an oxygen atom through a double bond.

carboxylic acid: an organic molecule containing –COOH as a functional group.

catalyst: a substance added to a chemical reaction that speeds up the reaction by altering the reaction mechanism and lowering the energy of activation; although it speeds up the reaction, it is never used up and is fully recovered.

cathode: electrode in an electrochemical cell where reduction takes place.

cation: a positively charged ion formed by the loss of electrons by an atom.

Charles's law: relationship of gases where the volume is directly proportional to the absolute temperature (Kelvin).

chromatography: separation technique in which substances are separated based on their polarity.

coffee cup calorimeter: instrument consisting of a Styrofoam cup used to measure the heat lost or gained during a chemical process.

colligative property: physical property of a solution that depends on the number of solute particles dissolved in the solution (boiling point elevation, freezing point depression, osmotic pressure, vapor pressure lowering).

common ion effect: shift of equilibrium induced by adding a common ion to the equilibrium, such as adding an acetate ion to a solution of acetic acid.

compound: grouping of two or more atoms bonded together.

concentration: the amount of solute in a given volume of solvent.

condensation: physical change in which a gas converts to a liquid.

conjugate acid: species produced after accepting a hydrogen ion from an acid.

conjugate base: species remaining after removing a hydrogen ion from an acid.

cooling curve: graph depicting the phase changes that occur as a substance is cooled.

coordination complex: compound containing a metal ion bonded to a group of surrounding molecules or ions that act as ligands.

Coulomb's law: electric force between two charged particles that are separated by a given distance.

covalent bond: chemical bond in which atoms share electrons; caused by the overlap of orbitals.

critical point: highest temperature and pressure where a gas and a liquid can co-exist in equilibrium.

crystalline solid: solid in which its internal arrangements of atoms or molecules repeats in a regular pattern.

current: the amount of charge passing a given point per second.

Dalton's law of partial pressures: if a container contains multiple gases, the total pressure inside the container is the sum of the pressures of the individual gases in the mixture.

delocalized electrons: electrons that are capable of moving to different locations in a molecule rather than remaining between two atoms.

dependent variable: variable in an experiment that is monitored and indicative of the change; plotted on the y-axis.

deposition: phase change in which a gas is converted directly into a solid.

diamagnetic: an atom whose electron configuration contains no unpaired electrons.

dipole moment: measure of the separation of positive and negative charges in a polar molecule.

dipole–dipole forces: intermolecular force between two polar molecules in close contact to one another.

disproportionation: a reaction in which one reactant undergoes both oxidation and reduction.

distillation: separation technique for liquids based on boiling points.

double bond: covalent bond involving two electron pairs shared between two atoms.

ductile: can be pulled into a wire.

effusion: escape of a gas through a small pinhole.

electrolyte: a solute that produces ions in aqueous solution and will conduct an electric current.

electrolytic cell: electrochemical cell in which energy must be added to the cell for oxidation and reduction to take place.

electron: negatively charged species located outside of the nucleus of an atom; has properties of both a wave and a particle.

electron affinity: the ability of an atom to draw electrons toward itself.

electron cloud: region in space where an electron can be located.

electron configuration: arrangement of electrons in the orbitals of an atom.

electronegativity: measure of the ability of an atom in a covalent bond to pull electrons toward itself.

electroplating: the deposition of a metal onto the surface of the cathode via a reduction half reaction.

element: pure substance that cannot be broken down further by chemical means.

empirical formula: formula that represents the smallest whole number ratio of atoms in a compound; may or may not be the same as the molecular formula.

endothermic: a chemical reaction in which the system absorbs energy from the surroundings (reaction feels cold).

energy levels: amount of energy associated with an atomic or molecular orbital.

enthalpy: heat content of a substance at constant pressure; energy released or absorbed during a chemical reaction.

entropy: a measure of the amount of randomness or disorder in a system.

equilibrium: ongoing and dynamic process in which a reaction forms products to a certain point then reforms the reactants.

ester: an organic molecule containing the –COO–C functional group within the molecule.

ether: an organic molecule containing C–O–C functional group within the molecule.

exothermic: a chemical reaction in which the system releases energy to the surroundings (reaction feels hot).

expanded octet: a Lewis structure in which the central atom contains more than eight electrons.

face-centered crystal: cubic unit cell that has lattice points at each corner as well as at the center of each side of the cube.

faraday: a unit of charge that equals the total charge of one mole of electrons.

filtration: technique for physical separation of a solid from a liquid by means of a filter.

fission: a nuclear reaction in which an atom is split.

formula unit: smallest whole number ratio between cations and anions in an ionic compound.

freezing point: temperature at which a liquid converts to a solid.

frequency: the number of cycles of a wave that pass a fixed point per unit of time.

fusion: a nuclear reaction in which the nuclei of two atoms combine.

gamma radiation: high energy electromagnetic radiation emitted during radioactive decay.

gas: matter that has no fixed volume or shape.

Gibbs free energy: measure of the amount of useful work a chemical reaction can do; determines whether a reaction will be spontaneous or nonspontaneous.

Graham's law of effusion: rate of effusion of a gas is inversely proportional to the square root of its molar mass.

gravimetric analysis: laboratory technique in which the mass of a substance can be stoichiometrically related to the mass of other substances.

ground state: lowest energy state of an atom or molecule.

group: vertical columns on the periodic table; chemical properties are similar to one another due to similar electron configurations.

half-life: time it takes for half of a substance to decay.

half-reaction: one of the two parts of an oxidation–reduction reaction representing either the oxidation or reduction of a substance.

halogen: group of reactive nonmetal elements located in Group VIIA on the periodic table.

heat: total kinetic energy of all of the particles in a sample.

heating curve: graph depicting the phase changes that occur as a substance is heated.

Henderson–Hasselbach: equation relating pH, pKa, and the concentrations of a weak acid and its conjugate base or a weak base and its conjugate acid.

Hess's law: heat involved for a chemical process can be expressed as the sum of individual reactions that, when added, equal the overall chemical reaction.

hybridization: mixing of different types of atomic orbitals to produce a set of equivalent hybridized orbitals.

Hund's rule: electrons having the same spin will occupy orbitals at the same energy before a pairing with opposite spin electrons occurs.

hydrate: an ionic compound that also contains water molecules trapped inside the crystal.

hydration: solvation when the solvent is water.

hydrogen bond: intermolecular force that occurs when a hydrogen atom is covalently bonded to either a fluorine, oxygen, or nitrogen (FON) atom and is simultaneously attracted to a neighboring nonmetal atom; strongest of all intermolecular forces.

hydrolysis: a reaction with water.

hydrophobic: water repelling.

hydrophilic: water attracting.

ideal gas: hypothetical gas whose temperature, pressure, and volume obey the ideal gas law.

ideal gas law: $PV = nRT$.

ideal solution: solution that obeys Raoult's law.

independent variable: variable that is changed in an experiment; plotted on x-axis.

indicator: compound that changes color based on the pH of the solution; used in acid–base titrations to determine the equivalence point.

inorganic compound: compound that is composed of elements other than carbon; of a non-biological origin.

insoluble: does not dissolve or dissociate in solution; precipitate.

ion: atom that has an overall charge due to the loss or gain of electrons.

intermediate: compound that is formed in one step of a reaction mechanism that is consumed in a later step.

ionic attraction: electrostatic attraction between cations and anions, governed by Coulomb's law.

ionic bond: bond that involves the transfer of electrons between atoms to form cations and anions that were held together by electrostatic attractions.

ionic crystals: lattice structures composed of repeating formula units of ionic compounds and held together by electrostatic attractions in a crystal lattice.

ionization energy: energy required to remove an electron in the ground state from an atom in the gas phase.

isoelectronic: atoms or ions that have the same electron configuration due to the same number of electrons.

isomers: compounds composed of the same elements that have different structural formulas.

isotopes: atoms of the same element that have different numbers of neutrons resulting in different mass numbers.

joule: SI unit of energy; $1 \text{ kg m} / \text{s}^2$.

ketone: an organic molecule containing the R–CO–R functional group.

law of conservation of energy: energy cannot be created or destroyed during chemical reactions, only changed from one form into another; nuclear reactions violate this principle.

law of conservation of mass: matter is neither created nor destroyed during a chemical reaction—atoms are merely rearranged to form new substances.

law of mass action: dictates that equilibrium constants/expressions are a ratio of the products in an equilibrium divided by the reactants, all of which are raised to the corresponding coefficients in the chemical equation.

Le Chatelier's principle: when stress is placed on a system at equilibrium (by changing the temperature, concentration, or pressure), the system will respond by shifting away from the stress.

Lewis structure: graphic representation of the covalent bonding between atoms in a molecule, where bonds are shown as lines and unpaired electrons are pairs of dots.

Lewis theory: theory of acids and bases in which acids are electron-pair acceptors and bases are electron-pair donors.

ligand: ion or molecule that forms a coordinate covalent bond to metal atoms or ions, resulting in a coordination complex.

limiting reactant: reactant in a chemical reaction that is entirely used up.

liquid: matter that has a fixed volume but takes the shape of its container.

magnetic quantum number: the third number in a set of quantum numbers that denotes the number and orientation of orbitals in a subshell.

magnetic spin quantum number: the fourth number in a set of quantum numbers that denotes the spin of an electron in an orbital.

mass number: the sum of the protons and neutrons in an atom giving rise to the overall mass.

mass percentage: percentage based on the mass of a component compared to the mass of the whole mixture for a solution.

malleable: can be hammered into a sheet or flattened.

mechanism: sequence of elementary steps or chemical reactions that describe an overall chemical change.

melting point: temperature at which a solid to liquid phase change occurs.

metal: element that is typically a solid at room temperature, has luster, conducts electricity and heat, and is malleable and ductile; on the left side of the periodic table; most elements are classified as metals.

metallic bonding: metallic nuclei are encompassed by a sea of electrons that are relatively mobile, giving rise to metallic properties.

miscible: two liquids that can be mixed to form a solution based on similar intermolecular forces between them.

molar mass: the number of grams in one mole of substance; can be found by using the atomic masses on the periodic table.

molality: concentration unit that is expressed by moles of solute divided by kilograms of solvent.

molarity: concentration unit that is expressed by moles of solute divided by liters of solution.

mole: a unit used to describe the amount of substance present in exactly 12 g of carbon-12; equivalent to 6.022×10^{23} particles.

molecular formula: formula that shows the exact number of atoms of each element present in a molecule or formula unit.

molecular orbital: orbitals created from atomic or hybridized orbitals.

molecule: compound that contains two or more atoms covalently bonded to one another.

net ionic equation: chemical equation that showcases the particles involved in the production of a product, such as a precipitate, gas, or liquid.

network covalent crystals: lattice structures created by an extensive network of covalent bonds, often containing the same type of atom.

neutralization: reaction between an acid and a base that produces water and a salt.

neutron: subatomic particle located in the nucleus of an atom that has no charge, but a mass that is essentially the same as a proton (but slightly heavier).

noble gas: elements located in Group VIIIA on the periodic table that have filled electron shells and are especially stable, unreactive elements.

nonelectrolyte: compound that does not conduct electricity when dissolved in a solution; covalent molecule.

nonideal solution: a solution that does not obey Raoult's law.

nonmetals: elements that are dull, brittle, and typically gases at room temperature and are poor conductors of heat and electricity; generally located on the right side of the periodic table.

normal boiling point: boiling point at 1 atm of pressure.

normal melting point: melting point at 1 atm of pressure.

nucleus: the massive center of an atom that contains protons and neutrons.

octet rule: atoms gain, lose, or share electrons in order to have eight valence electrons (filled *s*- and *p*-shells), thus acquiring a noble gas electron configuration.

orbital: three-dimensional region around the nucleus of an atom that describes the probable location of an electron.

orbital notation: picture representation of the distribution of electrons in the various orbitals for an atom or ion.

organic compound: compound containing carbon.

organic functional group: groups of atoms that have characteristic properties in carbon-containing compounds.

osmotic pressure: pressure required to inhibit the movement a solvent particles across a semi-permeable membrane.

oxidation: loss of electrons from an atom, thereby increasing the charge.

oxidation number: positive or negative charge on an atom.

oxidizing agent: the substance in a reduction-oxidation reaction that is reduced.

paramagnetic: an atom whose electron configuration contains one or more unpaired electrons.

partial pressure: pressure exerted by a component gas in a mixture of gases.

Pauli exclusion principle: there is a maximum of two electrons in any orbital and they must have opposite spins; no two electrons can have the same set of four quantum numbers.

period: horizontal rows on the periodic table.

pH: scale that describes the acidity of substances where acidic substances have values between 0 and 7 and bases are between 7 and 14; negative logarithm of the hydrogen ion (hydronium) concentration.

phase diagram: shows the states of matter of a substance at given temperatures and pressures.

photon: a quantum of light or electromagnetic radiation that has particulate properties but no mass.

physical properties: properties that can be measured without changing the identity of a substance.

principal quantum number: the first of four quantum numbers that describe the energy level of an electron.

pi bond: bond resulting from the overlap of unhybridized orbitals; additional bond between two atoms that is formed following the formation of a sigma bond.

polar covalent bond: bond in which the electrons are unevenly shared between atoms, creating separate centers of positive and negative charge; gives rise to dipole–dipole and hydrogen bonding of intermolecular forces.

precipitate: insoluble solid that is formed during a chemical reaction.

proton: subatomic particle located in the nucleus of an atom that has a positive charge; the number of protons in the nucleus distinguishes atom types.

quantum numbers: a set of four numbers that describe the probable location of an electron in an atom or ion.

radioactive decay: unstable nuclei break apart to form separate isotopes, releasing radiation such as alpha and beta particles or gamma rays during the process.

rate law: mathematical relationship that describes the relationship between concentration of reactants and the overall rate.

Raoult's law: describes how the vapor pressure of a solvent is depressed when a solute dissolves in the solvent.

rate-determining step: the slow step in a reaction mechanism that determines the overall rate of a reaction.

reaction order: the exponent to which the concentration of a reactant is raised to in a rate law.

reaction rate: a measure of the change in concentration of a reactant or product over time.

reaction quotient: value calculated from concentrations plugged into an equilibrium expression to determine the direction of an equilibrium shift.

reducing agent: the substance in a reduction-oxidation reaction that is oxidized.

reduction: gain of electrons by an atom, thereby decreasing the charge.

resonance structure: alternative Lewis structure that has the same arrangement of atoms with a different distribution on bonds.

salt: ionic compound formed as the result of an acid–base reaction containing the cation from the base and the anion from the acid.

salt bridge: part of a voltaic cell, utilized to maintain electrical neutrality in half-cells.

saturated solution: solution in which no additional solute can dissolve at a given temperature and pressure; specific for a given solute.

sigma bond: covalent bond resulting from the overlap of orbitals along the internuclear axis.

simple cubic unit cell: cubic unit cell that has lattice points at each corner.

single bond: covalent bond containing two electrons shared between two atoms.

solid: matter that has a fixed volume and shape.

solubility: the maximum amount of solute that will dissolve at a given temperature and pressure.

solute: the substance being dissolved in a solution; usually present in the smaller amount.

solution: homogeneous mixture of a solute dissolved in a solvent.

solvation: surrounding solute particles with solvent molecules.

solvent: the substance doing the dissolving in a solution; usually present in the larger amount.

specific heat: the amount of energy required to raise 1 gram of material by 1 °C.

spectator ion: charged species that does not participate in a chemical reaction but is present to balance the charge of other ions in a solution; ions that are not included in the net ionic equation.

standard heat of formation: amount of energy required to form a compound from its elements under standard conditions.

standard state conditions: 1 atm of pressure and 298 K (25°C).

standard temperature and pressure (STP): 1 atm of pressure and 273 K (0 °C).

standardization: technique in which a titration with a known standard is performed to determine the exact concentration of a solution.

state function: property of a system that is only dependent on present conditions, not the path by which it took.

structural formula: picture representation that shows how the atoms in a compound are bonded to one another.

subatomic particles: particles smaller than atoms, such as protons, neutrons, and electrons.

sublimation: conversion of a solid directly into a gas.

supersaturated solution: solution that contains more than the maximum amount of solute dissolved at a given temperature and pressure; unstable solution that can be crystallized by adding a single crystal of undissolved solute.

surrounding: everything that is outside of a system.

system: whatever chemical reaction or process being studied in the universe.

temperature: measure of the average kinetic energy of particles.

titration: lab technique in which substances are added to one another until the endpoint, at which the reactants are in perfect stoichiometry with one another; makes use of an indicator.

transition elements: metallic elements that have varying oxidation states depending on the specific bonding interactions; found on the periodic table in the d- and f-blocks.

transition state: unstable, high-energy arrangement of atoms midway between reactants and products in an energy diagram.

triple bond: bond containing three pairs of electrons shared between two atoms.

triple point: temperature and pressure on a phase diagram at which a substance is in all three states of matter in an equilibrium.

unit cell: smallest group of atoms that maintain the symmetry of a crystal lattice.

universe: contains both the system and surroundings.

unsaturated solution: solution in which more solute can dissolve at a given temperature and pressure.

VSEPR: valance shell electron pair repulsion theory; used in conjunction with Lewis structures to determine the overall shape and geometry of a molecule.

valence electrons: electrons that are furthest from the nucleus and determine the properties of elements; electrons showcased by the noble gas (short-hand) electron configurations.

van der Waals forces: intermolecular force caused by the formation of an instantaneous dipole and whose strength is proportional to the number of electrons in the molecule; weakest of all intermolecular forces; also called London dispersion or simply dispersion forces.

vapor pressure: pressure exerted by a substance in the gas phase that is normally in a different state of matter at a given temperature and pressure (vapor).

vapor pressure curve: graph that expresses the pressure exerted by a volatile liquid at various temperatures.

voltaic cell: electrochemical cell that converts chemical energy into electrical energy via a redox reaction.

wavelength: the distance between identical points on a repeating wave.

Index

PERIODIC TABLE
Atomic Properties of the Elements

NIST
National Institute of Standards and Technology
U.S. Department of Commerce

Physics Laboratory
physics.nist.gov

Standard Reference Data
www.nist.gov/srd

Frequently used fundamental physical constants

For the most accurate values of these and other constants, visit physics.nist.gov/constants
1 second = 9 192 631 770 periods of radiation corresponding to the transition between the two hyperfine levels of the ground state of ^{133}Cs

speed of light in vacuum	c	299 792 458 m s^{-1}	(exact)
Planck constant	h	6.6261 x 10^{-34} J s	($\hbar = h/2\pi$)
elementary charge	e	1.6022 x 10^{-19} C	
electron mass	m_e	9.1094 x 10^{-31} kg	
	$m_e c^2$	0.5110 MeV	
proton mass	m_p	1.6726 x 10^{-27} kg	
fine-structure constant	α	1/137.036	
Rydberg constant	R_∞	10 973 732 m^{-1}	
	$R_\infty c$	3.289 842 x 10^{15} Hz	
	$R_\infty hc$	13.6057 eV	
Boltzmann constant	k	1.3807 x 10^{-23} J K^{-1}	

Legend: Solids · Liquids · Gases · Artificially Prepared

Key

Atomic Number	58
Symbol	Ce
Name	Cerium
Atomic Weight†	140.116
Ground-state Configuration	[Xe]4f5d6s^2
Ionization Energy (eV)	5.5387
Ground-state Level	$^1G_4$

Elements

Z	Symbol	Name	Atomic Weight†	Ground-state Config.	Ionization (eV)	Level
1	H	Hydrogen	1.00794	1s	13.5984	$^2S_{1/2}$
2	He	Helium	4.002602	1s^2	24.5874	$^1S_0$
3	Li	Lithium	6.941	[He]2s	5.3917	$^2S_{1/2}$
4	Be	Beryllium	9.012182	[He]2s^2	9.3227	$^1S_0$
5	B	Boron	10.811	[He]2s^{2}2p	8.2980	$^2P_{1/2}$
6	C	Carbon	12.0107	[He]2s^{2}2p^2	11.2603	$^3P_0$
7	N	Nitrogen	14.0067	[He]2s^{2}2p^3	14.5341	$^4S_{3/2}$
8	O	Oxygen	15.9994	[He]2s^{2}2p^4	13.6181	$^3P_2$
9	F	Fluorine	18.9984032	[He]2s^{2}2p^5	17.4228	$^2P_{3/2}$
10	Ne	Neon	20.1797	[He]2s^{2}2p^6	21.5645	$^1S_0$
11	Na	Sodium	22.98976928	[Ne]3s	5.1391	$^2S_{1/2}$
12	Mg	Magnesium	24.3050	[Ne]3s^2	7.6462	$^1S_0$
13	Al	Aluminum	26.9815386	[Ne]3s^{2}3p	5.9858	$^2P_{1/2}$
14	Si	Silicon	28.0855	[Ne]3s^{2}3p^2	8.1517	$^3P_0$
15	P	Phosphorus	30.973762	[Ne]3s^{2}3p^3	10.4867	$^4S_{3/2}$
16	S	Sulfur	32.065	[Ne]3s^{2}3p^4	10.3600	$^3P_2$
17	Cl	Chlorine	35.453	[Ne]3s^{2}3p^5	12.9676	$^2P_{3/2}$
18	Ar	Argon	39.948	[Ne]3s^{2}3p^6	15.7596	$^1S_0$
19	K	Potassium	39.0983	[Ar]4s	4.3407	$^2S_{1/2}$
20	Ca	Calcium	40.078	[Ar]4s^2	6.1132	$^1S_0$
21	Sc	Scandium	44.955912	[Ar]3d4s^2	6.5615	$^2D_{3/2}$
22	Ti	Titanium	47.867	[Ar]3d^{2}4s^2	6.8281	$^3F_2$
23	V	Vanadium	50.9415	[Ar]3d^{3}4s^2	6.7462	$^4F_{3/2}$
24	Cr	Chromium	51.9961	[Ar]3d^{5}4s	6.7665	$^7S_3$
25	Mn	Manganese	54.938045	[Ar]3d^{5}4s^2	7.4340	$^6S_{5/2}$
26	Fe	Iron	55.845	[Ar]3d^{6}4s^2	7.9024	$^5D_4$
27	Co	Cobalt	58.933195	[Ar]3d^{7}4s^2	7.8810	$^4F_{9/2}$
28	Ni	Nickel	58.6934	[Ar]3d^{8}4s^2	7.6399	$^3F_4$
29	Cu	Copper	63.546	[Ar]3d^{10}4s	7.7264	$^2S_{1/2}$
30	Zn	Zinc	65.38	[Ar]3d^{10}4s^2	9.3942	$^1S_0$
31	Ga	Gallium	69.723	[Ar]3d^{10}4s^{2}4p	5.9993	$^2P_{1/2}$
32	Ge	Germanium	72.64	[Ar]3d^{10}4s^{2}4p^2	7.8994	$^3P_0$
33	As	Arsenic	74.92160	[Ar]3d^{10}4s^{2}4p^3	9.7886	$^4S_{3/2}$
34	Se	Selenium	78.96	[Ar]3d^{10}4s^{2}4p^4	9.7524	$^3P_2$
35	Br	Bromine	79.904	[Ar]3d^{10}4s^{2}4p^5	11.8138	$^2P_{3/2}$
36	Kr	Krypton	83.798	[Ar]3d^{10}4s^{2}4p^6	13.9996	$^1S_0$
37	Rb	Rubidium	85.4678	[Kr]5s	4.1771	$^2S_{1/2}$
38	Sr	Strontium	87.62	[Kr]5s^2	5.6949	$^1S_0$
39	Y	Yttrium	88.90585	[Kr]4d5s^2	6.2173	$^2D_{3/2}$
40	Zr	Zirconium	91.224	[Kr]4d^{2}5s^2	6.6339	$^3F_2$
41	Nb	Niobium	92.90638	[Kr]4d^{4}5s	6.7589	$^6D_{1/2}$
42	Mo	Molybdenum	95.96	[Kr]4d^{5}5s	7.0924	$^7S_3$
43	Tc	Technetium	(98)	[Kr]4d^{5}5s^2	7.28	$^6S_{5/2}$
44	Ru	Ruthenium	101.07	[Kr]4d^{7}5s	7.3605	$^5F_5$
45	Rh	Rhodium	102.90550	[Kr]4d^{8}5s	7.4589	$^4F_{9/2}$
46	Pd	Palladium	106.42	[Kr]4d^{10}	8.3369	$^1S_0$
47	Ag	Silver	107.8682	[Kr]4d^{10}5s	7.5762	$^2S_{1/2}$
48	Cd	Cadmium	112.411	[Kr]4d^{10}5s^2	8.9938	$^1S_0$
49	In	Indium	114.818	[Kr]4d^{10}5s^{2}5p	5.7864	$^2P_{1/2}$
50	Sn	Tin	118.710	[Kr]4d^{10}5s^{2}5p^2	7.3439	$^3P_0$
51	Sb	Antimony	121.760	[Kr]4d^{10}5s^{2}5p^3	8.6084	$^4S_{3/2}$
52	Te	Tellurium	127.60	[Kr]4d^{10}5s^{2}5p^4	9.0096	$^3P_2$
53	I	Iodine	126.90447	[Kr]4d^{10}5s^{2}5p^5	10.4513	$^2P_{3/2}$
54	Xe	Xenon	131.293	[Kr]4d^{10}5s^{2}5p^6	12.1298	$^1S_0$
55	Cs	Cesium	132.9054519	[Xe]6s	3.8939	$^2S_{1/2}$
56	Ba	Barium	137.327	[Xe]6s^2	5.2117	$^1S_0$
57	La	Lanthanum	138.90547	[Xe]5d6s^2	5.5769	$^2D_{3/2}$
58	Ce	Cerium	140.116	[Xe]4f5d6s^2	5.5387	$^1G_4$
59	Pr	Praseodymium	140.90765	[Xe]4f^{3}6s^2	5.473	$^4I_{9/2}$
60	Nd	Neodymium	144.242	[Xe]4f^{4}6s^2	5.5250	$^5I_4$
61	Pm	Promethium	(145)	[Xe]4f^{5}6s^2	5.582	$^6H_{5/2}$
62	Sm	Samarium	150.36	[Xe]4f^{6}6s^2	5.6437	$^7F_0$
63	Eu	Europium	151.964	[Xe]4f^{7}6s^2	5.6704	$^8S_{7/2}$
64	Gd	Gadolinium	157.25	[Xe]4f^{7}5d6s^2	6.1498	$^9D_2$
65	Tb	Terbium	158.92535	[Xe]4f^{9}6s^2	5.8638	$^6H_{15/2}$
66	Dy	Dysprosium	162.500	[Xe]4f^{10}6s^2	5.9389	$^5I_8$
67	Ho	Holmium	164.93032	[Xe]4f^{11}6s^2	6.0215	$^4I_{15/2}$
68	Er	Erbium	167.259	[Xe]4f^{12}6s^2	6.1077	$^3H_6$
69	Tm	Thulium	168.93421	[Xe]4f^{13}6s^2	6.1843	$^2F_{7/2}$
70	Yb	Ytterbium	173.054	[Xe]4f^{14}6s^2	6.2542	$^1S_0$
71	Lu	Lutetium	174.9668	[Xe]4f^{14}5d6s^2	5.4259	$^2D_{3/2}$
72	Hf	Hafnium	178.49	[Xe]4f^{14}5d^{2}6s^2	6.8251	$^3F_2$
73	Ta	Tantalum	180.94788	[Xe]4f^{14}5d^{3}6s^2	7.5496	$^4F_{3/2}$
74	W	Tungsten	183.84	[Xe]4f^{14}5d^{4}6s^2	7.8640	$^5D_0$
75	Re	Rhenium	186.207	[Xe]4f^{14}5d^{5}6s^2	7.8335	$^6S_{5/2}$
76	Os	Osmium	190.23	[Xe]4f^{14}5d^{6}6s^2	8.4382	$^5D_4$
77	Ir	Iridium	192.217	[Xe]4f^{14}5d^{7}6s^2	8.9670	$^4F_{9/2}$
78	Pt	Platinum	195.084	[Xe]4f^{14}5d^{9}6s	8.9588	$^3D_3$
79	Au	Gold	196.966569	[Xe]4f^{14}5d^{10}6s	9.2255	$^2S_{1/2}$
80	Hg	Mercury	200.59	[Xe]4f^{14}5d^{10}6s^2	10.4375	$^1S_0$
81	Tl	Thallium	204.3833	[Hg]6p	6.1082	$^2P_{1/2}$
82	Pb	Lead	207.2	[Hg]6p^2	7.4167	$^3P_0$
83	Bi	Bismuth	208.98040	[Hg]6p^3	7.2855	$^4S_{3/2}$
84	Po	Polonium	(209)	[Hg]6p^4	8.414	$^3P_2$
85	At	Astatine	(210)	[Hg]6p^5	—	$^2P_{3/2}$
86	Rn	Radon	(222)	[Hg]6p^6	10.7485	$^1S_0$
87	Fr	Francium	(223)	[Rn]7s	4.0727	$^2S_{1/2}$
88	Ra	Radium	(226)	[Rn]7s^2	5.2784	$^1S_0$
89	Ac	Actinium	(227)	[Rn]6d7s^2	5.3807	$^2D_{3/2}$
90	Th	Thorium	232.03806	[Rn]6d^{2}7s^2	6.3067	$^3F_2$
91	Pa	Protactinium	231.03588	[Rn]5f^{2}6d7s^2	5.89	$^4K_{11/2}$
92	U	Uranium	238.02891	[Rn]5f^{3}6d7s^2	6.1939	$^5L_6$
93	Np	Neptunium	(237)	[Rn]5f^{4}6d7s^2	6.2657	$^6L_{11/2}$
94	Pu	Plutonium	(244)	[Rn]5f^{6}7s^2	6.0260	$^7F_0$
95	Am	Americium	(243)	[Rn]5f^{7}7s^2	5.9738	$^8S_{7/2}$
96	Cm	Curium	(247)	[Rn]5f^{7}6d7s^2	5.9914	$^9D_2$
97	Bk	Berkelium	(247)	[Rn]5f^{9}7s^2	6.1979	$^6H_{15/2}$
98	Cf	Californium	(251)	[Rn]5f^{10}7s^2	6.2817	$^5I_8$
99	Es	Einsteinium	(252)	[Rn]5f^{11}7s^2	6.3676	$^4I_{15/2}$
100	Fm	Fermium	(257)	[Rn]5f^{12}7s^2	6.50	$^3H_6$
101	Md	Mendelevium	(258)	[Rn]5f^{13}7s^2	6.58	$^2F_{7/2}$
102	No	Nobelium	(259)	[Rn]5f^{14}7s^2	6.65	$^1S_0$
103	Lr	Lawrencium	(262)	[Rn]5f^{14}7s^{2}7p?	4.9?	$^2P_{1/2}?$
104	Rf	Rutherfordium	(265)	[Rn]5f^{14}6d^{2}7s^2	6.07	$^3F_2?$
105	Db	Dubnium	(268)			
106	Sg	Seaborgium	(271)			
107	Bh	Bohrium	(272)			
108	Hs	Hassium	(277)			
109	Mt	Meitnerium	(276)			
110	Ds	Darmstadtium	(281)			
111	Rg	Roentgenium	(280)			
112	Cn	Copernicium	(285)			
113	Uut	Ununtrium	(284)			
114	Uuq	Ununquadium	(289)			
115	Uup	Ununpentium	(288)			
116	Uuh	Ununhexium	(293)			
117	Uus	Ununseptium	(294)			
118	Uuo	Ununoctium	(294)			

†Based upon ^{12}C. ( ) indicates the mass number of the longest-lived isotope.

For a description of the data, visit physics.nist.gov/data

NIST SP 966 (September 2010)

Source: *National Institute of Standards and Technology*

Notes